Introduction to Data Multicasting

Lawrence Harte

Althos Publishing
Fuquay-Varina, NC 27526 USA
Telephone: 1-800-227-9681
Fax: 1-919-557-2261
email: info@althos.com
web: www.Althos.com

Althos

First Printing

Printed and Bound by Lightning Source, TN.

International Standard Book Number: 1-932813-55-1

About the Author

Mr. Harte is the president of Althos, an expert information provider which researches, trains, and publishes on technology and business industries. He has over 29 years of technology analysis, development, implementation, and business management experience. Mr. Harte has worked for leading companies including Ericsson/General Electric, Audiovox/Toshiba and Westinghouse and has consulted for hundreds of other companies. Mr. Harte continually researches, analyzes, and tests new communication technologies, applications, and services. He has authored over 100 books on telecommunications technologies and business systems covering topics such as mobile telephone systems, data communications, voice over data networks, broadband, prepaid services, billing systems, sales, and Internet marketing. Mr. Harte holds many degrees and certificates including an Executive MBA from Wake Forest University (1995) and a BSET from the University of the State of New York, (1990).

Table of Contents

Data Multicasting

Data multicasting is the process of transmitting media channels to a number of users through the use of multiple distributed channels.

Multicasting typically involves the use of group addresses, which allow receivers to "tune" to the same stream of data as it is transmitted over the network. This is in contrast to a unicast transmission whereby multiple copies of the stream are individually addressed to an end user and are transmitted over the network. Multicasting provides much more efficient use of the network resources. However, the use of multicasting can complicate media control functions such as pause and fast-forward.

Multicasting (one-to-many or many-to-many) can dramatically increase the efficiency of a network compared to unicasting (one-to-one) or broadcasting (one-to-all) transmission. Multicasting is critical for mass media streaming sources such as IP television and Internet radio. Without the use of multicasting, a 3 Mbps television streaming service would require data connections of 30 Gbps to provide service to 10,000 customers.

Multicasting group membership defines how members find, join and disconnect from multicast sessions. Multicast transmission involves the use of special multicast addresses. It is possible to configure multicast systems to provide varying levels of quality of service for different multicast members.

There are many types of protocols (commands and processes) that can be used to setup and manage multicast sessions. Some of the key protocols used for multicasting include IGMP, PIM-DM, PIM-SM, MOSPF, CBT and BGMP. The processes and capabilities that these protocols have determine the amount of latency (setup and transmission delay), scalability (ability to serve many users) and protocol overhead (percentage of network resources that are needed for the protocol commands and operation).

Multicast sessions may use security processes to ensure administrators can configure multicast trees and only authorized members may attach and decode multicast media. There are other emerging forms of multicast transmission that include gridcasting and peercasting where multicast recipients retransmit media to other users.

Multicast Applications

Multicast applications are programs or functions that involve communication with one to many or many to many applications. They range from best effort distribution to many receivers (such as television services through the Internet), to high-reliability real time distribution (such as critical video conferencing services).

Content Distribution

Content distribution is the process of transferring content to one or more persons, companies or points. Content distribution (such as television programs) may need to be distributed in near-real time without the ability to recover lost media.

Shared Applications

Shared applications are processes or functions that allow multiple devices or users to interact with them. Examples of shared applications include multi-user word processor programs, whiteboards and multi-user gaming programs. Shared applications may require reliable near-real time distribution to allow participates to effectively interact with the applications.

Information Distribution

Multipoint information distribution applications are delivery services such as news, sports, stock or weather updates that transports data or media to groups of users. Information distribution tends to be less time sensitive.

Network Configuration

Network configuration is a set of network conditions and parameters that are used to allow one or more types of services and applications. Networks can be composed of many devices that require near simultaneous updates and multicasting can be used to update these devices close to the same time period. These types of services may require tracking or guaranteed delivery services.

Multipoint Distribution

Sending signals to multiple users requires the use of a combination of unicasting, broadcasting, anycasting or multicasting services that can perform multipoint distribution.

Unicasting

Unicasting is the process of transmitting media channels to a number of users through the use of a separate channel (unicast channel) for each user.

Each unicast channel can be separately setup, managed and disconnected under the control of the serving computer host. If unicasting is used to provide broadcast or multicast services, a separate communication session must be established and managed between each user (client) and the broadcast provider (media server).

Figure 1.1 shows how multipoint distribution can be performed using unicast transmission. Each recipient must create a separate unicast connection directly to the media source. This means as each connection (viewer) is added, the media server and the routers closest to the source must be able to simultaneously transport all connections (4 connections in this example).

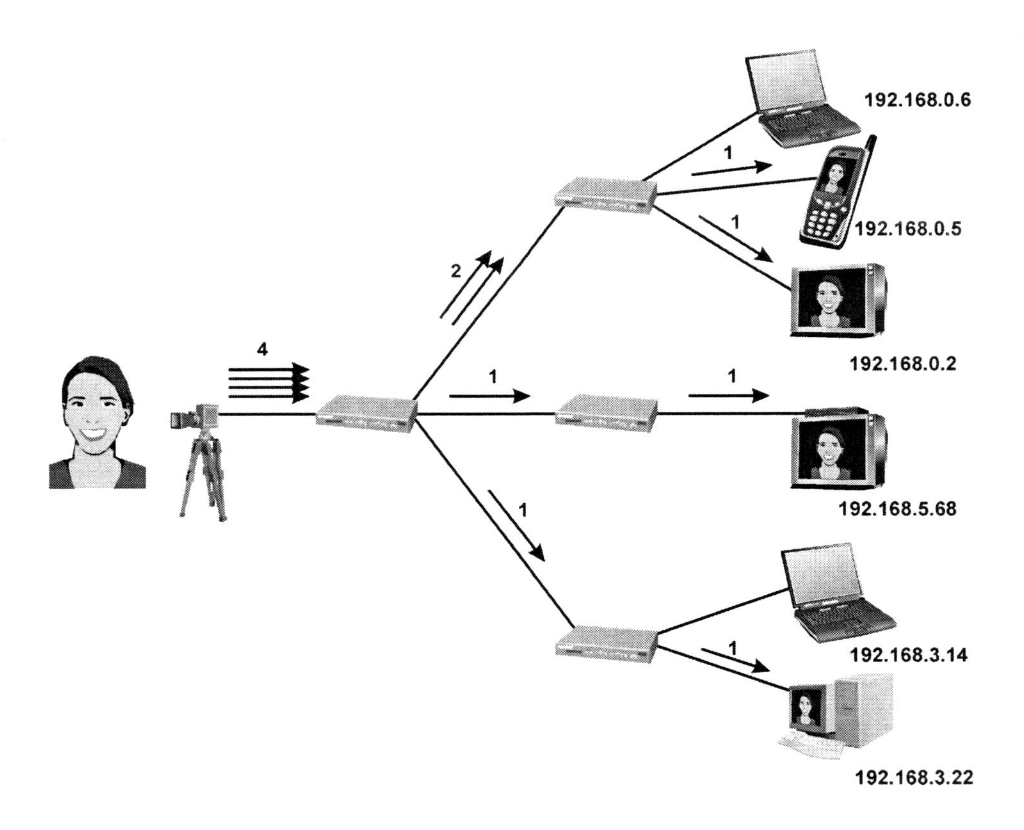

Figure 1.1, Multipoint Distribution Using Unicasting

Broadcasting

Broadcasting is a process that sends voice, data or video signals simultaneously to group of people or companies in a specific geographic area or who can connect or receive signals from a broadcast network system (e.g. satellite or cable television system). Broadcasting is typically associated with radio or television radio transmission systems that send the same radio signal to many receivers in a geographic area.

Broadcasting can also be applied to data distribution systems where all users that are connected to the network can receive and forward the same information signal. When a broadcast packet of data is received from an upstream source, it can be copied and distributed to all connections in a downstream direction. Broadcast transmission systems do not know or care about who is interested in receiving the packets so every packet is distributed to every receiver regardless if they desire to receive the data or not.

Figure 1.2 shows how multipoint distribution can be performed using broadcast transmission. Each router in the data network must receive packets from the upstream connection, copy each packet and the packets in the downstream direction. Each router in the data network receives packets, copies the packets and forwards the packets until they reach all recipients in the system.

Anycasting

Anycasting is the process of setting up media streaming connections to the best or closest source. While anycast distribution may allow all users to find and connect to a media source, the tree structure for an anycast system may not be the most efficient distribution structure.

Each router in the anycast data network must receive packets from any multicast stream, copy each packet and forward the packets in the downstream direction. Each recipient can find and connect to any path in the distribution tree.

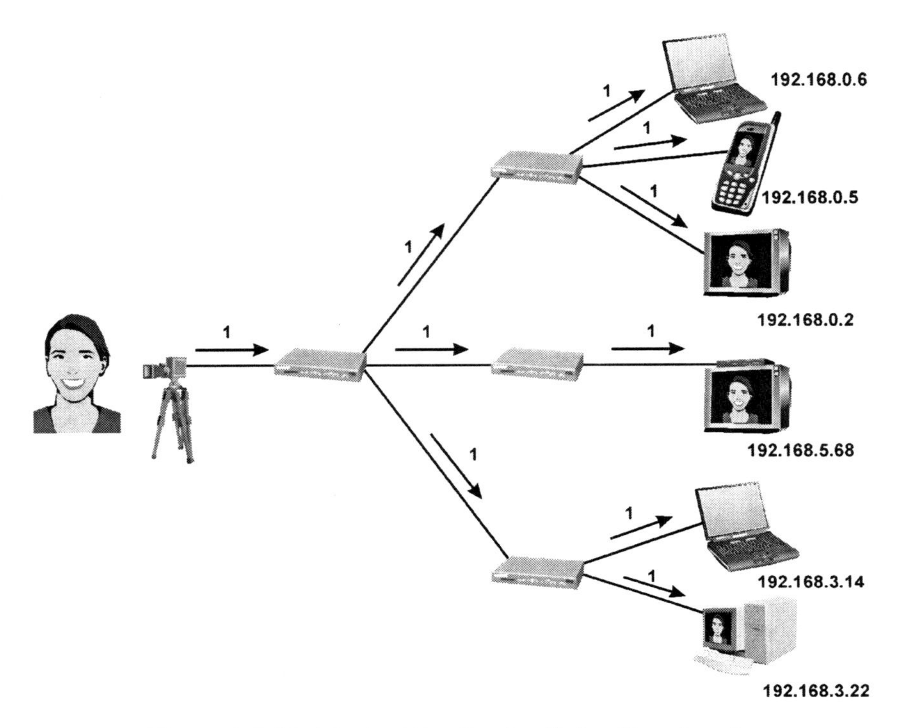

Figure 1.2, Multipoint Distribution Using Broadcasting

Figure 1.3 shows how multipoint distribution can be performed using anycast transmission. The recipient can find and request a connection to a multicast stream. This example shows that the connection may not be the ideal or shortest path between the source and the recipient.

Multicasting

Multicasting is the process of transmitting media channels to a number of users through the use of distributed channels (copying media channels) as they progress through a network. Using multicast medium access control (MAC) or Internet protocol (IP) addresses, multiple end users may "tune" to the same stream of data as it is transmitted over the network.

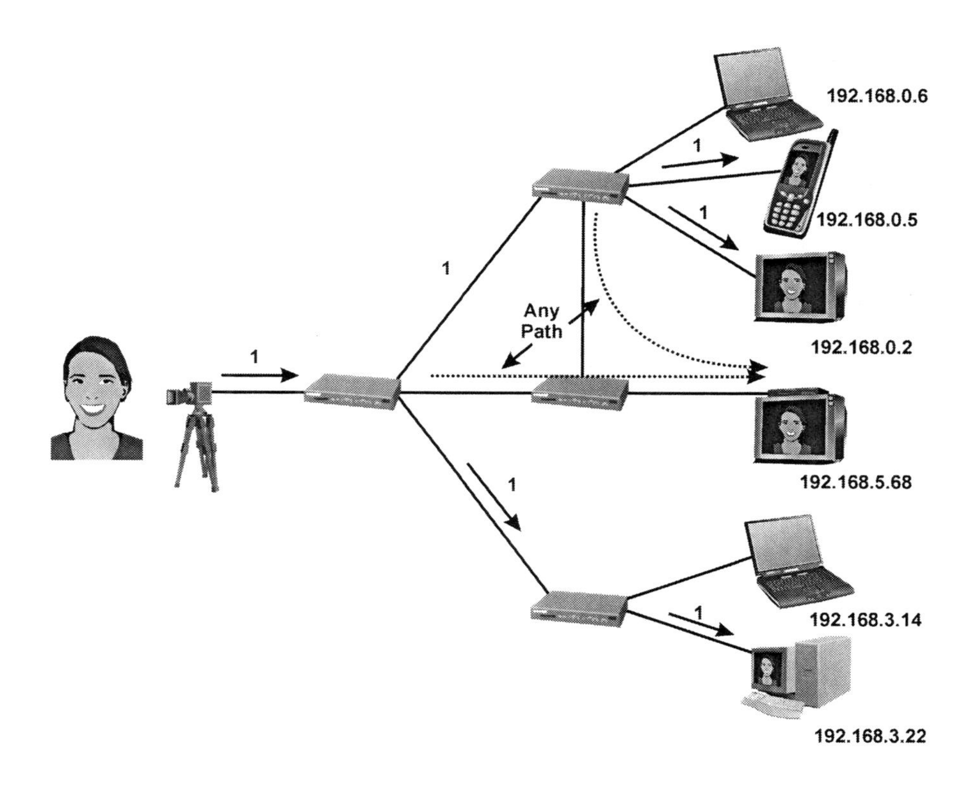

Figure 1.3, Multipoint Distribution Using Anycasting

Multicasting provides much more efficient use of the network resources because routers only receive, copy and forward packets towards their destination if there are recipients downstream of the router.

Figure 1.4 shows how multipoint distribution can be performed using multicast transmission. Each router in the data network that is part of the multicast distribution tree must receive packets from the upstream connection and copy each packet in the downstream direction. Each router in the data network only receives and forwards packets if there are multicast recipients downstream in the system.

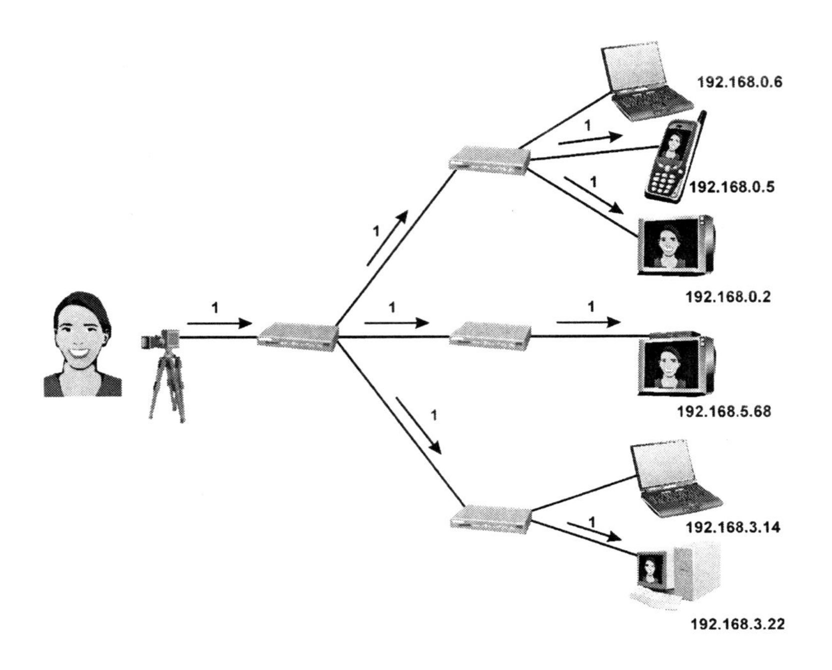

Figure 1.4, Multipoint Distribution Using Multicasting

Data Routing

Packet data routing involves the transmission of packets through intelligent switches (called routers) that analyze the destination address of the packet and determine a path that will help the packet travel towards its destination. The common types of data packets that are routed through a network include Internet protocol version 4 (IPv4) and IPv6.

Internet Protocol Version 4 (IPv4)

IP version 4 is a network packet routing protocol that uses 32 bit addresses. To help simplify the presentation of IPv4 addresses, it is common to group each 8 bit part of the IP address as a decimal number separated from the other parts by a dot(.), such as 207.169.222.45.

Internet Protocol Version 6 (IPv6)

IP version 6 is a network packet routing protocol that uses 128 bit addresses. IPV6 is an enhanced version of Internet protocol version (4 IPv4) that was developed primarily to correct shortcomings of IPv4 such as the 32 bit address that limited the maximum number of devices that could be addresses and to extend the capabilities of IP to meet the demands of the future such as data multicasting and improved quality of service (QoS) capabilities. IPv6 addresses are denoted as 8 hexadecimal numbers, separated by colons.

A typical address will look like this:

0800:5008:0000:0000:0000:1005:AABC:AD46

A short-hand notation that replaces one set of consecutive zeros with colons (::) may also be used. The above address can also be denoted by:

0800:5008::1005:AABC:AD46

A routing table contains at least two pairs of information; a destination address and a next hop routing address. Because of this, routers only need to hold information that they can use to determine the optimal paths to their destinations. Routers may also hold the cost of routing information that may be based on per packet cost or link usage cost.

Figure 1.5 shows how a router is used to receive and forward data packets toward their destination. This diagram shows that a router receives packets from other routers, examines the destination address of the packet and uses its stored routing tables to determine which router is the best choice for forwarding the packet towards its destination.

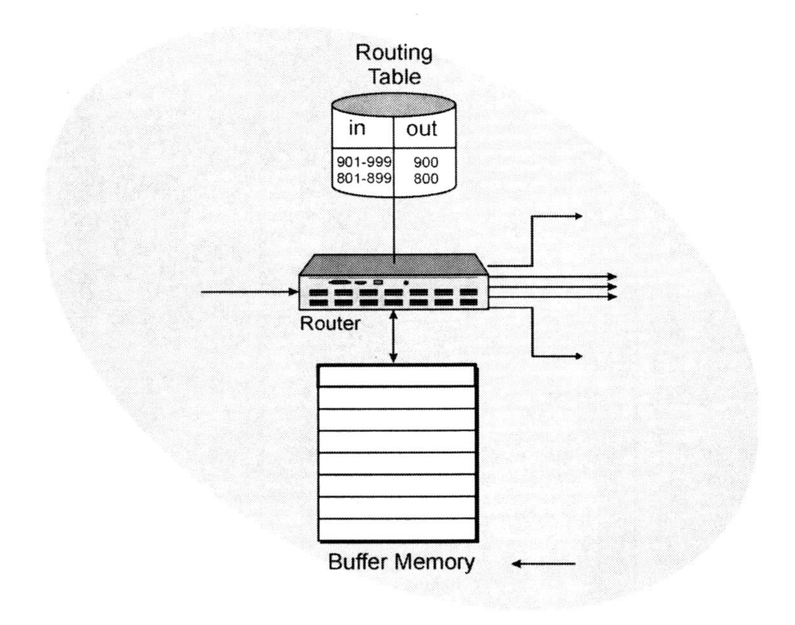

Figure 1.5, Data Routing

There are two key methods (routing algorithms) that routers can use to determine where to send packets; distance vector routing and link state routing.

Distance Vector Routing

Distance vector routing algorithms are packet processing rules that allow routers to discover, determine and use the distances of connection routes to make decisions on how to forward packets. In general, distance vector algorithms may require less processing power and memory than other routing protocols. Distance vector routing algorithms may be more prone to the creation of routing loops and they are less scalable and resilient than other routing algorithms.

A routing loop is a connection path or process that passes data packets from a router to other routers that return the packet back to the original sending router. This may occur because routing tables are dynamically changing as routers continually learn of connections to new routers, which results in the creation of new paths for forwarding packets. For packets that get caught in a routing loop, the time to live (TTL) field in the packet header will eventually be depleted (decreases each time a routing hop occurs) so the packet will eventually be discarded.

Link State Routing

Link state algorithms are packet processing rules that allow routers to discover, determine and use link connection routes to make decisions on how to forward packets. In general, link state algorithms are less prone to the creation of routing loops and they are more scalable and more resilient than other routing algorithms. However, link state algorithms may require more processing power and memory than other routing protocols.

Multicast Operation

Multicast systems setup and distribute media through the establishment of distribution trees. These trees can be setup by allowing users to request connection to the media source (sparse mode) or by flooding the network to all potential recipients and disconnecting the connections to users that are not interested in receiving the connection (dense mode).

Multicast Session

A multicast session is the connections and media transfers that occur between sources of information and group members that are attached to the multicast tree.

Sparse Mode Multicasting

Sparse mode multicasting is the distribution of media to multiple users within a data network where a limited or small number of the users that are connected to the network are part of the multicast group. Each member of a sparse mode multicasting system requests connection to the multicast group and this establishes the multicast tree.

Figure 1.6 shows how a data network may be setup to allow sparse mode multicasting. This example shows that sparse mode multicasting is effective when there are a small number of devices that want to receive a shared media source. Each device requests connection to the media source when they want to view it. Only the routers that become part of the distribution tree need to communicate with the group members.

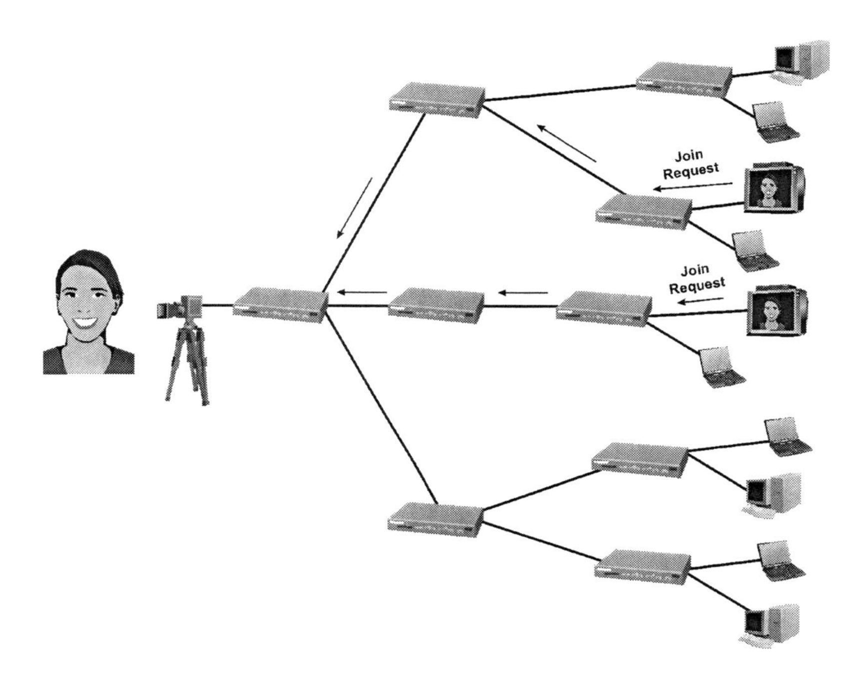

Figure 1.6, Sparse Mode Multicasting

Dense Mode Multicasting

Dense mode multicasting is the distribution of media to multiple users within a data network where many or most of the users that are connected to the network are part of the multicast group. In dense mode multicast systems, the network is flooded with multicast messages and group members who are not connected to the network are pruned from the multicast tree. Members of a dense mode multicasting system often request connections to the multicast group after the tree has been created.

Flooding

Flooding is an action of forwarding a frame onto all ports of a switch except the port on which it arrived. Flooding may be used for frames with multicast addresses or unknown unicast destination addresses.

Pruning

Pruning is the process of disconnecting connections that are no longer active in a communication network. An example of pruning is the removal of users in a dense mode multicast network that have been added but did not subscribe to or are not using the multicast service. Multicast pruning is a process that removes connections in a multicast tree that are no longer needed or used.

Grafting

Grafting is the process of adding a leaf or group member to a multicast session while the multicast session is in progress (late entry or reconnection). Grafting allows late entry to a multicast session.

Figure 1.7 shows how a data network may be setup to allow dense mode multicasting. This example shows that dense mode multicasting floods all routers in the network with multicast information. Devices that are not interested are removed (pruned) from the distribution tree.

Group Management

Group management is the process of defining groups of users or devices and adding and removing members (people and/or devices) to the groups. Group management protocols such as IGMP are used to add, manage and remove

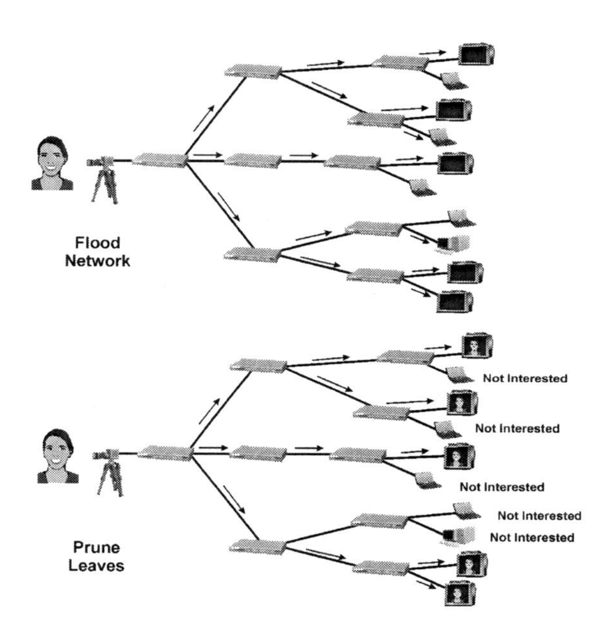

Figure 1.7, Dense Mode Multicasting

members in a group. A multicast group manager is a device or process that is responsible for maintaining a list and coordinating members of a group. Group membership protocols include Internet group management protocol (IGMP) and multicast listener discovery (MLD) protocol.

Group Addressing

A group address (multicast address) is an identification code that is used to identify a set of stations or a number of devices as the destination for transmitted data. In essence, devices in a multicast group listen for (share) a single address that identifies a multicast session. Multicast addresses can be dynamically assigned for specific multicast sessions.

The Internet assigned numbering authority is a group that is responsible for the assignment and coordination of Internet addresses and key parameters such as protocol variables and domain names. The IANA controls the assignment of IP multicast addresses.

Of the 32 bit Internet addresses version 4, 28 bits have been reserved as multicast addresses. These are known as class D IP addresses and the multicast IP addresses range from 224.0.0.0 to 239.255.255.255.

Of the 128 bit Internet addresses version 6, 112 bits have been reserved for multicast addresses. Of the 48 bit Ethernet addresses, 23 bits have been reserved for multicast addresses.

Figure 1.8 shows the different types of multicast group addressing. IP version 4 uses 28 bits of the 32 bit IPv4 address for multicast addresses. Ethernet multicast addressing uses 23 of the 48 bit Ethernet address for multicast addressing. IP version 6 uses 112 bits of the 132 bits for multicast addressing and some of the remaining bits are used for control purposes (flag and scope bits).

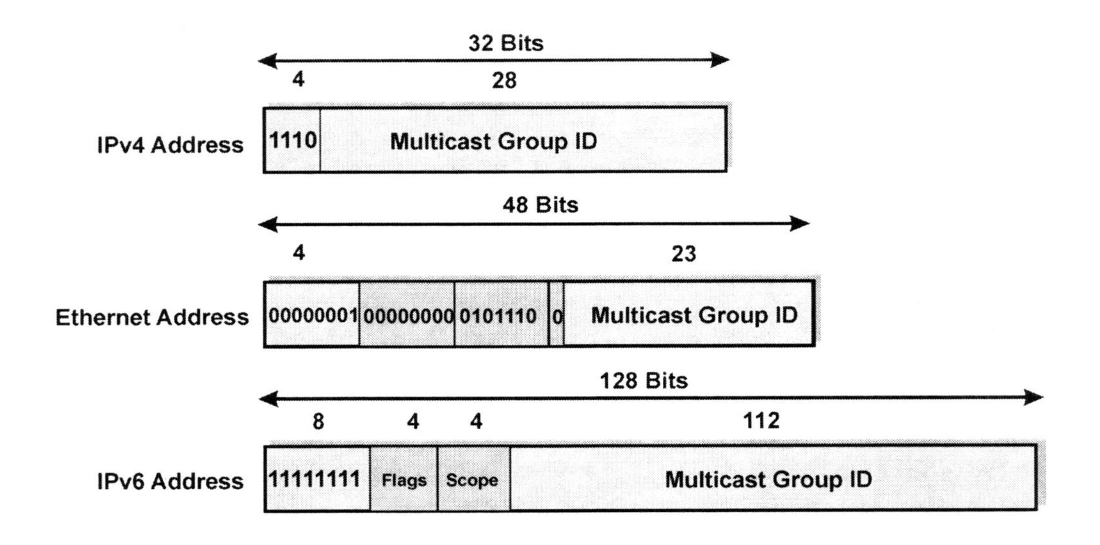

Figure 1.8, Multicast Group Addressing

Multicast Scoping

Multicast scoping is the process of assigning the number of routing hop constraints or using addresses that limit where data packets can travel through a network. An example of administrative scoping is the use of multicast addresses, which cannot be transmitted outside a local domain.

Address Scoping

Address scoping is the assigning or defining of a set of addresses that are used for specific functions or restricted to specific areas (within a domain or subnetwork).

Limited Scoped Address

A limited scoped address is a unique private IP address in the range of 239.0.0.0 to 239.255.255.255 that is associated with multicast connections. Limited scoped addresses can be used to allow a multicast session to be used in private networks without interference from or transfer to the public Internet.

Time to Live (TTL)

Time to live is a field within a data packet that is used to limit the maximum number of routing or switching points a packet may pass through (a hop limit) during transmission in a data network. The TTL counter is decreased as it progresses through each router or switching point in the network. If the TTL counter reaches 0, the packet can be discarded. The use of TTL ensures packets will not be transmitted in an infinite loop.

Distribution Trees

A multicast distribution tree is a path structure from a source to its destinations where the origination point (tree base) divides (branches) into its destination points (tree leaves). Distribution trees can have a single source or the branches on the tree may be used as other source feeds (shared tree).

Source Tree

A source tree is a multicast structure that has the source of the multicasting located at the base of the tree. A source tree is also known as a shortest path tree because each branch in the tree has the shortest path to other branches or nodes.

A single source multicast source tree has a single starting point and all of the nodes in the tree are the optimal (shortest) paths that can be traveled from the source to the receiver. This diagram shows that the addressing format is (S, G) where S is the source IP address and G is the multicast IP address.

Figure 1.9 shows how a single source multicast data session allows a single source to send the same information to multiple receivers without the need to repeat the transmission back through multiple switches and routers in the network. This example shows that an IP address source is combined with a single multicast address that allows each router in the multicast tree to forward the packets only to members of the group.

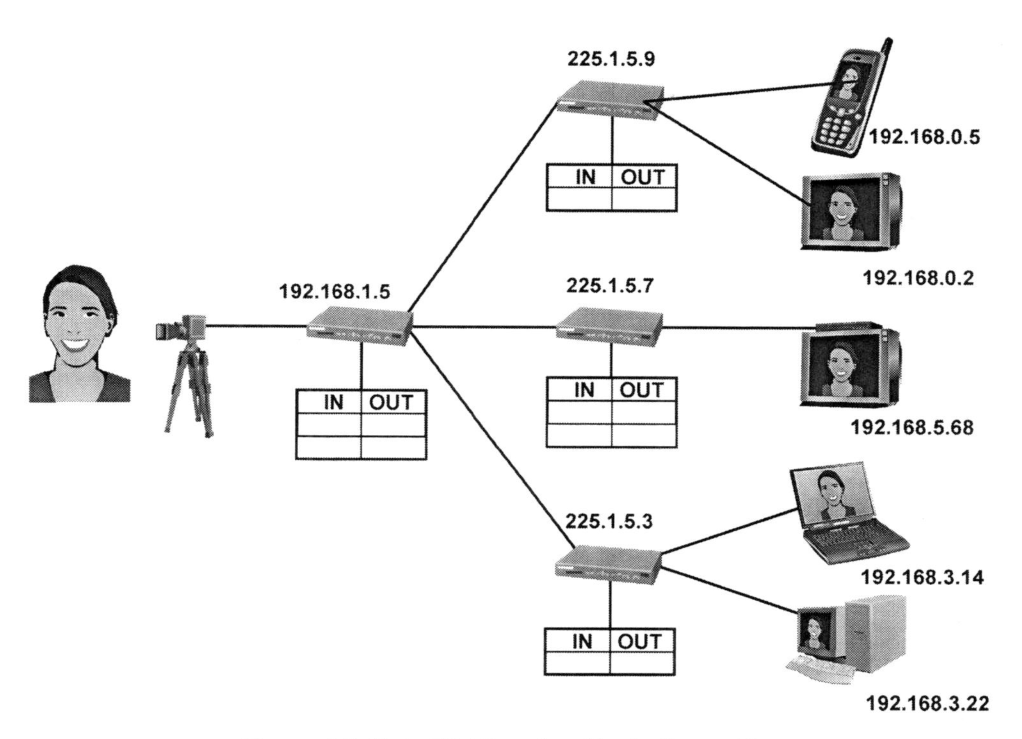

Figure 1.9, Data Multicasting Single Source Tree

Shared Tree

A shared tree is a multicast structure that has the source of the multicasting located anywhere in the tree. The location of the source in a shared tree is called the rendezvous point (RP).

Rendezvous Point (RP)

A rendezvous point is a data network address of a node (such as a router) in a communications network where one or multiple data sources will be directed to so they can be retransmitted to other points in a multicast distribution tree.

A multicast shared tree can have multiple sources that supply to a rendezvous point (RP) where is it distributed by a tree to the group receivers. This example shows that a shared tree structure may have redundant or less than optimal (longer) paths from a source to the receivers. This diagram shows that the addressing format is (*,G) where * is the source IP address (can be multiple sources) and G is the multicast IP address.

Figure 1.10 shows how a shared source multicast data session allows multiple sources to send the information to multiple receivers through the use of a reference address (rendezvous point). In this example, the data from each source is sent to the rendezvous point (RP), which then distributes the information through a tree structure to the multicast group recipients. This diagram shows that there is the potential for some duplicate transmission in the shared source multicast session as the source and destination may be sent in different directions through the same routers.

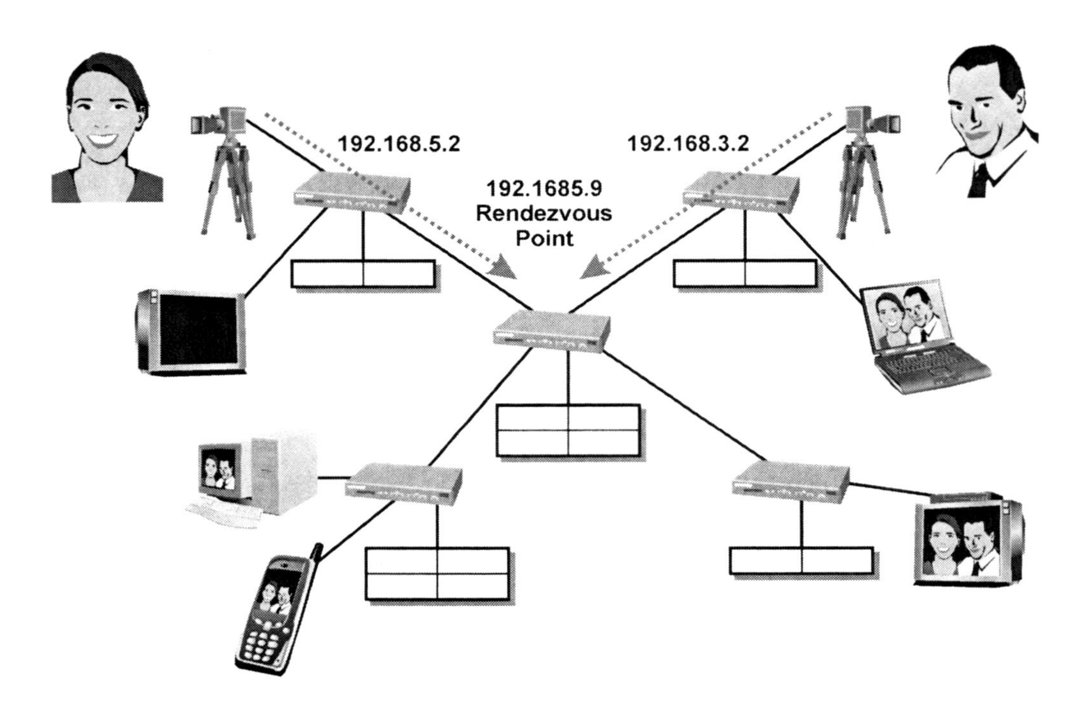

Figure 1.10, Data Multicast for Shared Sources

Multicast Address Notation

Multicast address notation is the format of addresses that indicate the multicast group and its source of information. For source trees, the notation is source and group pair (S, G) and for shared trees, the notation is any source and group pair (*, G). A source and group pair (S,G) is the identification information used for multicast source trees. The notation indicates the source address and the group address.

A source and group pair (S, G) is the identification information used for multicast source trees. For multicast shared trees that can have sources at any location in the tree, the notation is changed to (*, G), which indicates the source address (any address) and the group address.

Tree Building

Tree building is the process that is used to create a setup of multicast connections that link the source(s) to the multicast group members. Multicast trees can be dynamically created as users are added to a multicast group (data driven) or they may be pre-setup before group members begin to participate (control driven).

Trees are setup to ensure loop free distribution. If multicast trees were not established, forwarding paths may be setup to allow data to loop from the source back to the source. Loop free is a network that contains connections that transfer information to locations different than its source location (the connection path does not return to the source).

Data Driven Tree Building

Data driven tree building is the process of creating the tree or branches of the tree as new members are discovered or when join requests are received. Examples of data driven tree building include distance vector based trees or shortest path multicast trees.

Control Driven Tree Building

Control driven tree building is the process of establishing a tree before group members begin to participate in the multicast session. An example of control driven tree building is the use of a core based tree (CBT) to establish a multicast distribution backbone between multiple multicast distribution domains.

Volatile Tree Structure

A volatile tree structure is a set of connection paths in a multicast tree that continually change over time.

Route Flapping

Route flapping is the continual changing of a network connection path that results from an intermittent congestion or loss of circuit connection that indicates to the current router connection path that there is a loss in connection or that a better connection path exists. This causes the packet routing path to continually change. These different paths can result in significant variance in transmission delay times (excessive jitter). Router flapping is overcome by newer IPV6 protocols and reservation protocols.

Core Based Trees (CBT)

Core based trees are a distribution structure for sparse mode multicasting in a data network. For CBT systems, new group members send join messages to a designated core router. Any routers that the join message passes along the way also begin to relay the multicast join request message. This builds a tree up as the packets move to the core. CBT is described in RFC 2189.

Core based trees are a distribution structure for sparse mode multicasting in a data network. For CBT systems, new group members send join messages to a designated core router. Any routers that the join message passes along the way also begins to relay the multicast. This builds a tree up as the packets move to the core.

CBT systems require that routers store tree state information, which limits the ability of CBT trees to be used (scaled) for large distribution systems. Because routers store CBT tree information, CBT trees can be bidirectional trees allowing data to flow in either direction between the source and the members in the tree.

Because explicit tree routing information is stored in each router in a CBT tree, the amount of processing that is required by the CBT router can be small when compared to other multicast protocols which require path calculations each time a multicast packet is received.

Figure 1.11 shows how a core based tree is setup. This diagram shows that group members send their join request to adjacent routers which forward the request up to the root node (core) of the CBT tree. The core routers are used to distribute multicast streams to non-core routers which forward the data to other non-core routers which deliver the data to multicast group members.

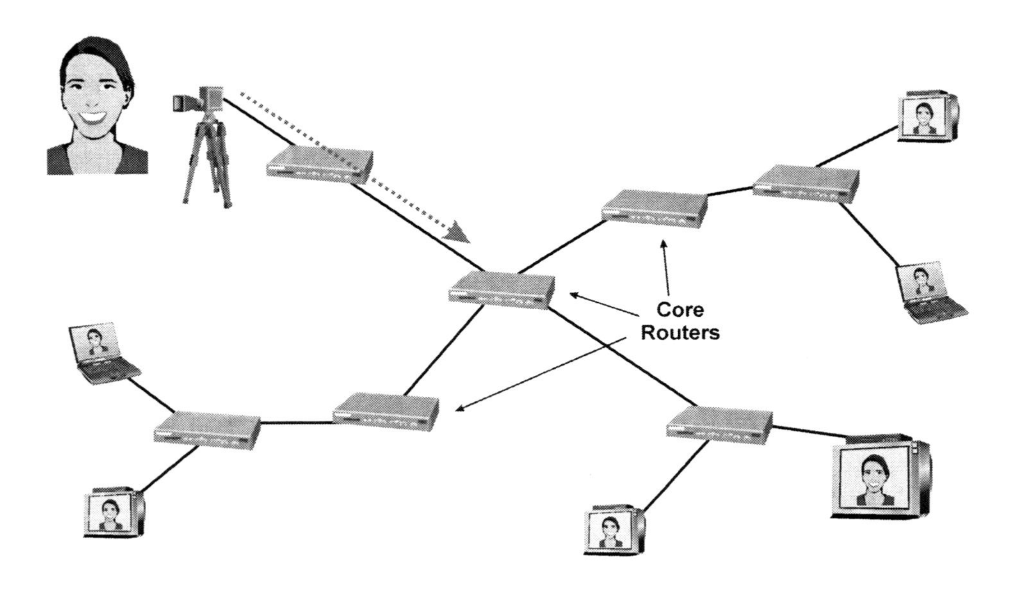

Figure 1.11, Core Based Tree

Join Styles

Join styles are the processes that are used to add members to a multicast group. The join process may be explicit or implicit.

Explicit Joins

Explicit joins require the group member to acknowledge or initiate communication with the multicast router or group manager to join a multicast session

Implicit Joins

Implicit joins allow the group member to receive and process multicast sessions without the need to acknowledge or communicate with the multicast router or group manager to participate in a multicast session.

Multicast Routing Table

A multicast routing table is a list (a database table) that is located within a router capable of multicasting that is used to determine which multicast packets will be copied and what addresses they should be forwarding to.

Multicast Forwarding

Multicast forwarding is the process of receiving data packets, checking multicast routing tables to determine if the packet should be forwarded and coping and sending data packets to paths or destinations that are identified in the multicast routing tables.

Reverse Path Forwarding (RPF)

Reverse path forwarding is the process of forwarding the transmission of data packets to paths that are downstream (in a forward direction) in a data communication system. Reverse path forwarding operates by reviewing the source and destination addresses of the data packet to its routing table to determine if the destination address is not located between the source and the current router address. If the address is not between the source and the current router (in the forward path), the packet can be forwarded towards its destination.

Reverse path forwarding check is the process of comparing the source of a packet and the incoming port or path to determine if the packet has been sent from a router that is upstream (between the router and the source) or downstream (from a router that is after the router). RPF is used to only accept multicast packet requests from upstream routers to avoid the possibility of data communication loops.

Forwarding loop is a sequence of packet transmission through routers that eventually brings it back to its originating points. A time to live (TTL) counter can be used in a data packet that discards the packet after a number of forwarding steps to disable the potential of forwarding loops.

Figure 1.12 shows how a reverse path forwarding check is used to prevent the potential for data communication loops in multicast networks. This example shows a data communication network that has multiple media servers (M1 and M2) that are connected to the data network. Media server 1 has initiated a multicast session and routers in the network receive a mul-

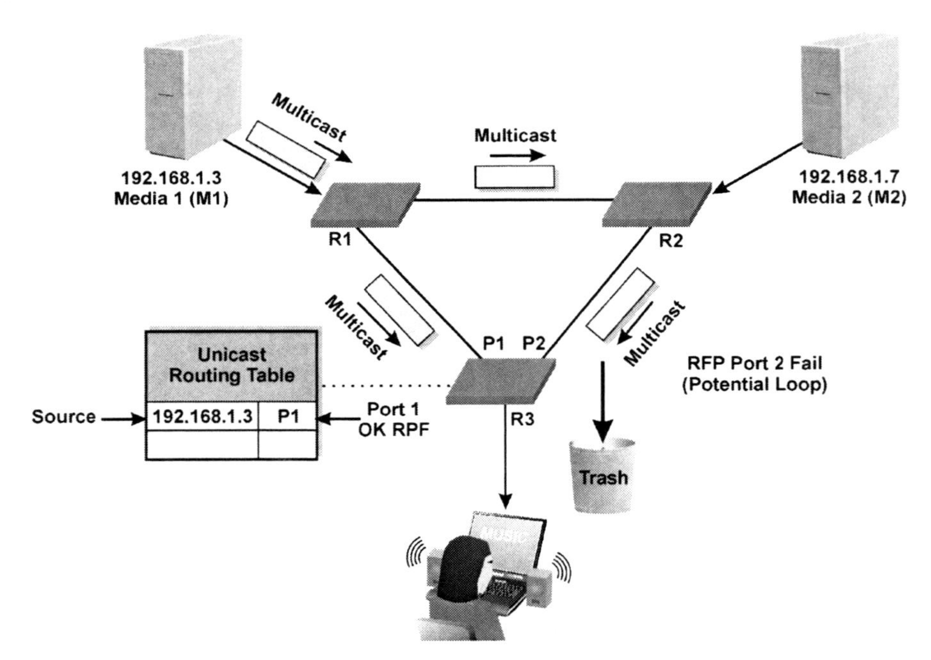

Figure 1.12, Reverse Path Forwarding Check

ticast request packet. As the multicast request packet is received by each router, the multicast routing table is checked to determine if the packet was received from a source that is upstream (correct path) or downstream. If the packet is received on a port that is downstream, the packet is discarded.

Ethernet Multicasting

Ethernet multicasting is the process of assigning a MAC address from a range of Ethernet MAC addresses dedicated for multicast sessions that allow devices to receive and capture multicast data that is distributed through an Ethernet network. Ethernet devices that are part of a multicast group will detect and receive packets that contain the multicast MAC address. Each Ethernet device has a unique Ethernet hardware address.

Multicast Addresses Mapping

Multicast address mapping is a process of translating multicast group addresses to the addresses of specific devices within a network (such as Ethernet addresses). When incoming multicast data packets are received, the multicast mapping process identifies the destination addresses and readdresses the packets for the destination devices.

Flat Address Space

Flat address space is a set of addresses that cannot be reassigned for other purposes (such as subnetting).

Address Filters

Address filtering is the process of receiving packets and reviewing if these packets should be processed or allowed to transfer through a device.

Address Ambiguities

Address ambiguities are the ability of an identification label to refer to more than one device or service.

Figure 1.13 shows how IP multicast addresses can be translated (mapped) into Ethernet multicast addresses. This example shows that of the 32 bit IP address, 28 bits are used for multicast addresses. For the 48 bit Ethernet multicast address, only 23 bits can be transferred from the IP address to the Ethernet address.

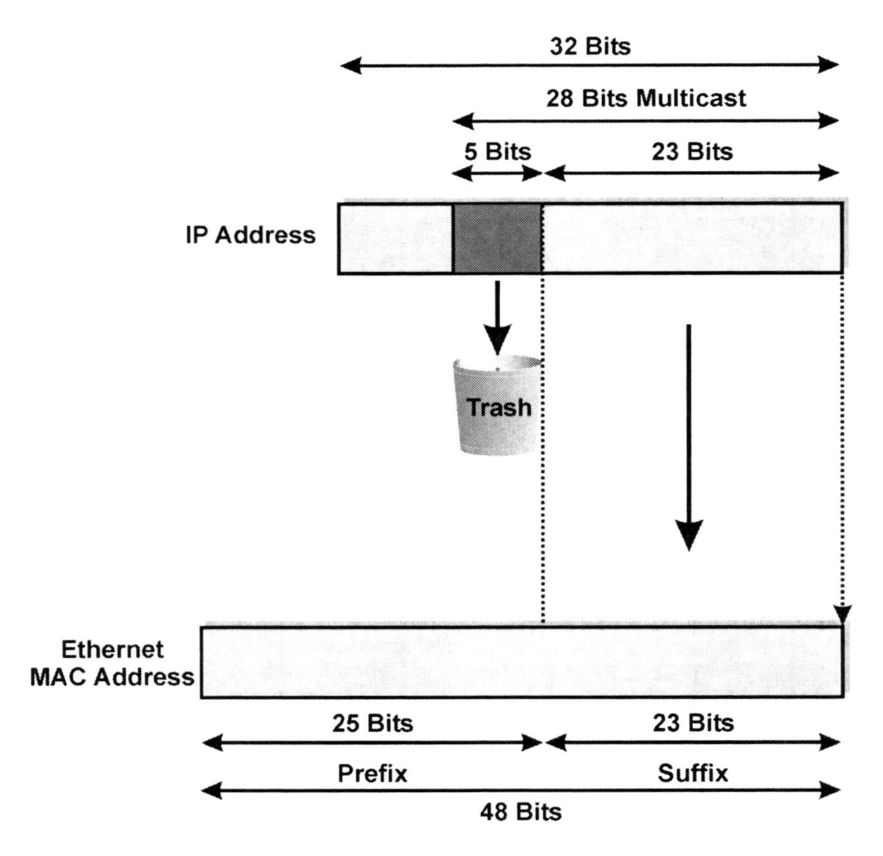

Figure 1.13 Mapping IP Multicast to Ethernet Multicast Addresses

Bandwidth Reservation

Reserved bandwidth is the allocation of a portion of transmission or processing capability of a device or system for specific users, purposes or applications. Bandwidth allocation can be reserved and controlled in a device using protocols such as resource reservation protocol (RSVP).

Resource Reservation Protocol (RSVP)

Resource reservation protocol is a set of commands and processes that can be used to control the quality of service (QoS) for packet flow for specific communication sessions in a packet data network. RSVP is primarily used in real-time communication sessions such as video or voice services over packet data networks.

Multicast Protocols

Multicast protocols are the languages, processes, and procedures that perform functions used to send control messages and coordinate the adding (setup), management or removal of multicast sessions. Multicast groups receive signals or information that is distributed content to multiple recipients or users. Examples of multicast protocol include IGMP and PIM.

Internet Group Management Protocol (IGMP)

Internet group management protocol is the commands, processes and procedures that are used to send control messages and coordinate the multicasting (simultaneous distribution) of data through an Internet protocol network. IGMP is used to establish membership into a multicast group that is operating within a network. Using IGMP, users can inform routers within the network that they would like to receive media and control messages from a specific multicast group. IGMP is defined in RFCs 1112 and 2236.

Key IGMP messages include a membership query and membership report. A membership query is a request for a device or system to identify the users or devices that are part of their membership (e.g. multicast) group. Multicast routers periodically send (typically every 1 to 3 minutes) them to group members to determine if they are still interested in participating in the multicast session. A membership report is a list of users or devices that are part of their membership (e.g. multicast) group.

Figure 1.14 shows how Internet group management protocol is used to setup and manage multicast communication sessions to a local multicast host.

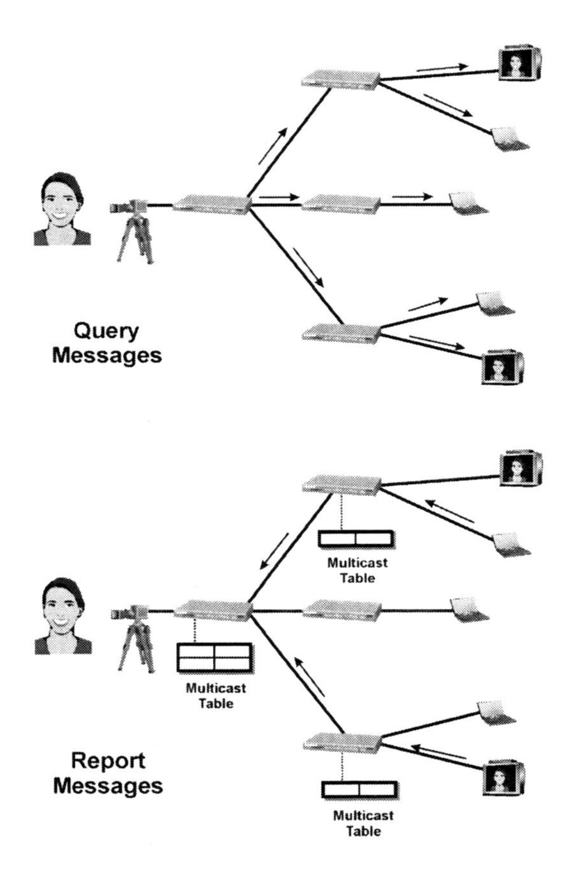

Figure 1.14, Internet Group Management Protocol (IGMP)

This example shows that the IGMP manager sends a query requests to group members to determine if they are interested (or still interested) in participating in a multicast session. Group members send reports to identify that they are interested in participating in a multicast session. The multicast routers use the report messages to build their multicast router tables.

IGMPv1

Internet group management protocol version 1 is the initial version of IGMP released in 1989 based on RFC 1112. IGMP sends membership queries every 60 seconds.

Figure 1.15 shows the format of an IGMP packet, version 1. IGMPv1 packets contain a type classification which identifies the packet as an IGMP report (join request) or an IGMP query (is the device still interested in receiving multicast packets).

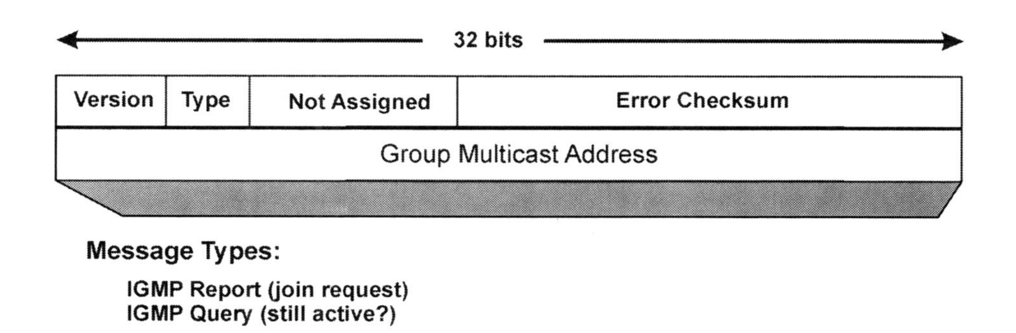

Figure 1.15, IGMP Packet Version 1

IGMPv2

Internet group management protocol version 2 is the second version of IGMP released in 1997 based on RFC 2236. IGMPv2 added the capability of devices to send leave messages.

Figure 1.16 shows the format of an IGMP packet, version 2. IGMPv2 packets contain a type classification which identifies the packet as an IGMP report (join request), IGMP leave (disconnection request), or an IGMP query (is the device still interested in receiving multicast packets). IGMPv2 messages also identify the maximum response time for the IGMP request.

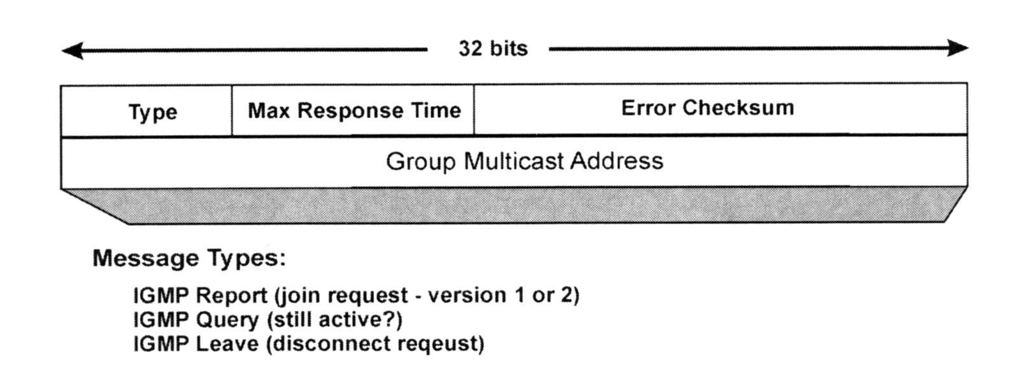

Figure 1.16, IGMP Packet Version 2

IGMPv3

Internet group management protocol version 3 is the third version of IGMP released in 2002 based on RFC 3376. IGMPv3 was a significant revision to the previous versions. Some of the key features of IGMPv3 include the ability of hosts to use list to identify devices they are willing to receive traffic from and the ability to block packets that come from sources identified as senders of unwanted traffic.

IGMP enables source specific multicasting which allows recipients to directly connect to the source of a multicast session rather than receiving multicast signals through other distribution points (rendezvous points). IGMPv3 includes include and exclude modes of operation.

Include mode is a condition of a system or method of operation where items or commands that are sent are used to add (include) additional services or options. An example of include mode is the use of group management messages in a multicast system to identify additional multicast streams or services that are to be added to a receiver.

Exclude mode is a condition of a system or method of operation where items or commands that are sent are used to remove or block additional services or options. An example of exclude mode is the use of group management messages in a multicast system to identify additional multicast streams or services that are not to be forwarded or received.

Figure 1.17 shows the format of an IGMP packet, version 3. IGMPv3 packets contain additional fields that can be used to define multicast sessions to

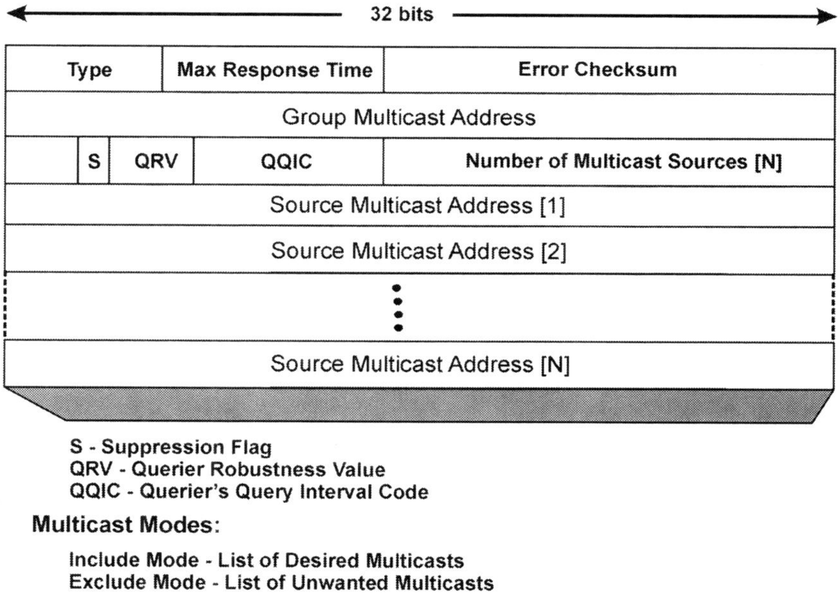

32 bits

Type | Max Response Time | Error Checksum
Group Multicast Address
S | QRV | QQIC | Number of Multicast Sources [N]
Source Multicast Address [1]
Source Multicast Address [2]
Source Multicast Address [N]

S - Suppression Flag
QRV - Querier Robustness Value
QQIC - Querier's Query Interval Code

Multicast Modes:

Include Mode - List of Desired Multicasts
Exclude Mode - List of Unwanted Multicasts

Figure 1.17, IGMP Packet Version 3

include to or exclude from its connection. This diagram shows that IBMPv3 packets can contain a set of include and exclude messages in the IGMP data packet.

Internet Group Management Protocol Snooping (IGMP Snooping)

Internet group management protocol snooping is the process of looking inside packets for IGMP messages so that the router can update its multicast routing tables as group members are added or removed from the multicast distribution tree. Internet Group Management Protocol (IGMP) Snooping enables DSLAMs, PON Optical Line Terminals (OLTs) and routers to passively monitor subscriber traffic in order to identify and properly assign multicast group membership. Access platforms incorporating this feature check IGMP packets passing through, pick out the group registration information and configure multicasting accordingly. Via IGMP snooping, multicast group traffic is only forwarded to ports servicing members identified as belonging to that particular multicast group. IGMP snooping generates no additional network traffic, allowing carriers to reduce network congestion.

Figure 1.18 shows how routers perform IGMP snooping. This example shows that a router has IGMP snooping software that allows it to decode the contents of each packet as it passes through the router. The router stores the packet, decodes the packet header and determines if the packet contains an IGMP message. If so, the router can use the information in the IGMP message to update its multicast routing table.

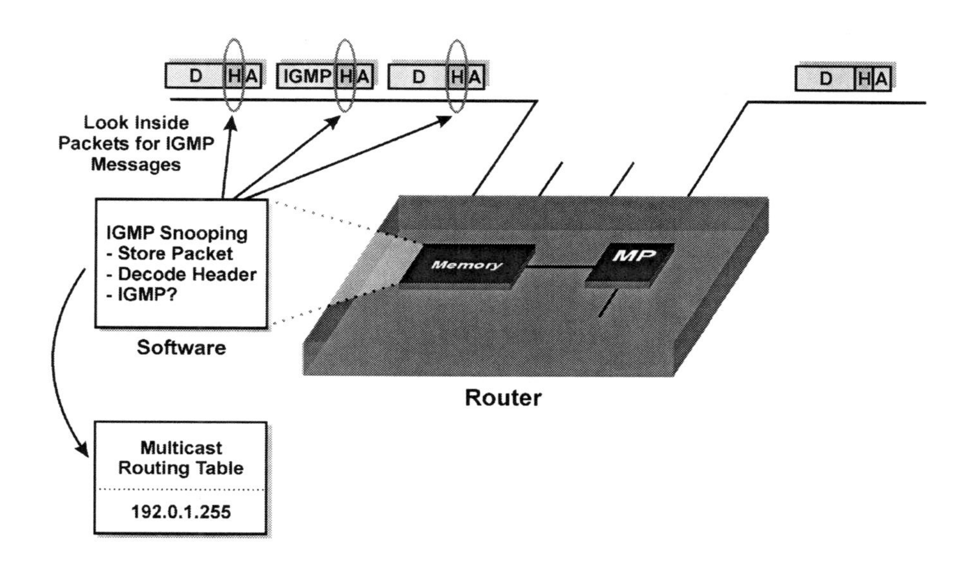

Figure 1.18, IGMP Snooping

Protocol Independent Multicasting (PIM)

Protocol independent multicasting is the setup and distribution of packets in a multicast tree structure using protocols (such as OSPG, static routes or BGP) other than multicast protocols. Protocol independent multicasting uses existing routing tables to determine if packets should be copied and routed in a multicast tree. This allows protocol independent multicasting (PIM) to be implemented on networks that use different underlying protocols.

As multicast packets are received at the router, their existing routing tables are used to determine if the address that the packets originated from would be sent back to their source. If so, the packet has arrived from a node that is upstream to the tree and the packet should be copied and forwarded to other ports on the router.

Protocol Independent Multicasting Sparse Mode (PIM-SM)

Protocol independent multicast-sparse mode is a data pull process that builds a tree to all the potential group recipients through the use of group join messages. PIM-SM can use any underlying unicast protocol to determine when new tree branches should be added. PIM-SM offers the most efficient method of multicasting with the disadvantage of longer delays in setting up the network.

Protocol Independent Multicasting Dense Mode (PIM-DM)

Protocol independent multicast-dense mode is a data push process that periodically floods a data network to build a tree to all the potential group recipients and then remove (prune) unwanted receivers in the tree. PIM-DM can use any underlying unicast protocol to determine how the tree should be constructed. The flood process may be continually repeated every 3 minutes to ensure all new group members can connect to the tree. PIM-DM offers the quickest method of setting up a multicast tree with the disadvantage of increased network data congestion.

Bidirectional Protocol Independent Multicasting Dense Mode (Bidir-PIM)

Bidirectional protocol independent multicasting is an enhanced version of protocol independent multicasting that allows one or more providers of content to send and receive data or media with one or more multicast receivers (many to many). Bidir-PIM modifies the PIM protocol to enable and efficiently transfer data in both directions.

Source Specific Multicast (SSM)

Source specific multicast is an extension to the protocol independent multicasting protocol that allows a multicast group member to define which multicast sources they want or don't want to connect to.

Border Gateway Multicast Protocol (BGMP)

Border gateway multicast protocol is a protocol that is used to communicate between routers to help determine the selection of routes for multicast packets over multiple network domains.

Multicast Listener Discover (MLD)

Multicast listener discover is the commands, processes and procedures that are used to send control messages and coordinate the multicasting (simultaneous distribution) of data through an Internet protocol network version 6. MLD is used to establish membership into a multicast group that is operating within a network. Using MLD, users can inform routers within the network that they would like to receive media and control messages from a specific multicast group.

Broadcast Media Distribution Protocol (BMDP)

Broadcast media distribution protocol are the commands and processes that control and enable broadcast distribution systems (such as IPTV) to adjust storage and distribution heuristically according to trends in subscriber behavior. BMDP also provides a buffer against network jitter and greater tolerance for peak bursts in traffic. And example of how BDMP operates is the setup of temporary storage of a popular digital television program as it is streamed through a multicast distribution system so that viewers can watch the program from a local media server rather than having to reconnect to a distant media server.

Distance Vector Multicast Routing Protocol (DVMRP)

Distance vector multicast routing protocol are the commands and processes that assist a router in the selection of routes based on the distance between connections. DVMRP keeps track of the distance of other connections between routers in a network. DVMRP uses Internet Group-Management Protocol (IGMP) to transfer routing information with neighboring routers.

A distance vector is a measurement of the length of a path between points and the direction the path has (from its origin to its end). When used in networks, a distance vector may represent the length of a connection or the number of routers between connection points.

Distance vector routing algorithms are packet processing rules that allow routers to discover, determine and use the distances of connection routes to make decisions on how to forward packets. In general, distance vector algorithms may require less processing power and memory than other routing protocols. Distance vector routing algorithms may be more prone to the creation of routing loops and they are less scalable and less resilient than other routing algorithms.

Multicast Extension to Open Shortest Path First (MOSPF)

Multicast extension to open shortest path first is a dense mode multicast protocol that uses a routing process that chooses the next unused shortest path when building a multicast tree. MOSPF routers make their forwarding decisions based on its forwarding cache which is created from a shortest path tree. A multicast forwarding cache is a temporary copy of a multicast forwarding table that is stored in a router.

Source Specific Multicast (SSM)

Source specific multicast is a protocol that enables a multicast recipient to request a connection directly to a multicast source rather than through a single rendezvous distribution point.

Any Source Multicast (ASM)

Any source multicast is a protocol that enables a multicast recipient to dynamically establish multicast connections to multicast trees.

Using ASM, a multicast group member indicates an interest in receiving traffic from a multicast address. The multicast network then discovers which multicast sources are capable of sending to that multicast address (multiple sources) and sets up connections between the multicast sources to that address.

ASM is well suited for groupware applications, which allow multiple people to simultaneously share participation in projects or data editing and the list of participants is not previously known and can dynamically change during the groupware session.

Source Filtering Group Membership Protocol (SGMP)

Source filtering group membership protocol is a set of commands and processes that can be used to allow for the exclusion (blocking) of sources in a multicast tree.

Cisco Group Management Protocol (CGMP)

Cisco group management protocol is a set of commands and processes that can be used to provide information to routers in a multicast tree about the additions (joins) and removals (leaves) that occur by group members. CGMP is used to remove or reduce the need to do IGMP snooping to find changes in multicast group membership.

Negative-Acknowledgment (NACK) Oriented Reliable Multicast (NORM)

Negative-acknowledgment (NACK) oriented reliable multicast is a set of commands and processes (a protocol) that are used to provide reliable transport on unmanaged IP networks. NORM uses a selective negative acknowledgment process that can be combined with additional protocol processes to provide coordinated transmission between senders and receivers.

Multicast Transport Protocol (MTP)

Multicast transport protocol is a set of commands and processes that are used to setup and coordinate the distribution of a source to multiple destinations using underlying network protocols to perform the transfer. MTP defines key member types including master, producers and consumers. The master coordinates the ordering of the messages, the producer sends the messages and the consumer receives the messages.

Router-Port Group Management Protocol (RGMP)

Router-port group management protocol is a set of commands and processes that are used to setup and manage multicast tree paths on a data network.

Pragmatic General Multicast (PGM)

Pragmatic general multicast is a transport protocol that provides reliable transmission of data on a multicast system. PGM operates on multicast transmitters and receivers to identify and control the successful delivery of data packets through a multicast tree distribution system.

Interdomain Multicast Protocols

Interdomain multicast protocols are commands and processes that can be used to advertise and setup multicast sessions between different domains. Interdomain multicast protocols are used to setup multicast distribution systems in different systems or at multiple geographic areas within a system.

Multiprotocol Border Gateway Protocol (MBGP)

Multiprotocol border gateway protocol are the commands and processes used by routers that are located between different networks to evaluate each of the possible multicast routes for the best one before choosing the routing path for multicast packets it receives and forwards. MBGP can be found in RFC 2283.

Multicast Source Discovery Protocol (MSDP)

Multicast source discovery protocol is a set of commands and processes that are used to allow rendezvous points (RPs) to transfer information about media sources between RPs in other domains (other networks).

Anycast RP

An anycast RP is an application of multicast networks (MSDP) that can be used to configure multicast routers for fault tolerance and load sharing.

Fault tolerance is the ability of a network or sub-system to continue to operate in the event of a hardware or software failure. Fault tolerant systems are typically able to identify the fault and replace the failed component or sub-system with another piece of equipment.

Load sharing is a process by which signaling traffic is distributed over two or more signaling or message routes to equalize and efficiently handle traffic for security purposes.

Multicast Listener Discovery (MLD)

Multicast listener discovery is an IPv6 protocol that can be used to send control messages and coordinate the multicasting (simultaneous distribution) of data through an IPv6. MLD is used to establish membership into a multicast group that is operating within a network. Using MLD, users can inform routers within the network that they would like to receive media and control messages from a specific multicast group. MLD is described in RFC 3810.

Truncated Reverse Path Broadcasting (TRPB)

Truncated reverse path broadcasting is the transmission of broadcast signals or packets to regions of the network, which have receivers (group members) that have indicated an interest in receiving the broadcast. TRPB may be accomplished by using IGMP packets to determine which branches or areas of a network have interested receivers and disabling or truncating the delivery of messages to areas that have no interested recipients.

Multicast Session Management

Multicast session management involves session description, session announcement, session initiation, and session control.

Session Description

Session description protocol (SDP) is a text based protocol that is used to provide high-level definitions of connections and media streams. The SDP protocol is used with session initiated protocol (SIP). The SDP protocol is used in a variety of communication systems including 3G wireless and the PacketCable system. SDP is defined in RFC 2327.

Session Announcement

Session announcement protocol is a set of commands and processes that are used to distribute information about communication sessions. SAP is defined in RFC 2974.

Session Initiation

Session initiation protocol is a set of text commands and processes that work with applications (application layer) to setup, manage and terminate communication sessions. SIP is a simplified version of the ITU H.323 packet multimedia system. SIP is defined in RFC 2543.

Session Control

Session control protocol is a set of commands and processes that are used to negotiate and initialize protocols that are used for communication sessions. An example of a session control protocol is session conference control protocol (SCCP).

Multicast Security

Multicast security is the ability of a system that distributes data or media to multiple people to maintain its desired well being or operation without damage, theft or compromise of its resources from unwanted people or events.

Identity Verification

Identity verification is a process of exchanging information that allows the confirmation of the true identity of the user (or device). Identify verification may be used to allow a service provider to decide to enable or deny service to users that cannot be identified.

Hash Function

A hash function is a mathematical process that converts (transforms) a block of data or information into an output value. Hash functions are one-way processes that result in the output value (called a hash or digest) not being convertible back into its original data or information form.

A message digest is a result of a calculation of a message (a hash) using a known key and algorithm.

Authentication

Authentication is a process of exchanging information between a communications device (typically a user device such as a mobile phone or computing device) and a communications network that allows the carrier or network operator to confirm the true identity of the user (or device). This validation of the authenticity of the user or device allows a service provider to deny service to users that cannot be identified. Thus, authentication inhibits fraudulent use of a communication device that does not contain the proper identification information.

Security association is the setup of a relationship between two network elements that ensures that traffic passing through the interface is authentic, unchanged and/or cryptographically secure (typically, through encryption.)

Non-Repudiation

Non-repudiation is the use of information or evidence that confirms that a person or company has received or used the products or services that they have ordered.

Encryption

Encryption is a process of a protecting voice or data information from being obtained by unauthorized users. Encryption involves the use of a data processing algorithm (formula program) that uses one or more secret keys that both the sender and receiver of the information use to encrypt and decrypt the information. Without the encryption algorithm and key(s), unauthorized listeners cannot decode the message. When the encryption and decryption keys are the same, the encryption process is known as symmetrical encryption. When different encryption and decryption keys are used (such as in a public encryption system), the process is known as asymmetrical encryption.

Public Key Encryption

Public key encryption is an authentication and encryption process that uses two keys (public key and a private key) to setup and perform encryption between communication devices. The public key and private keys can be combined to increase the key length provider and a more secure encryption system.

Group Key

A group key is a value that can be used by multiple devices or users to encrypt and decrypt messages.

Trusted Authority

A trusted authority is an information source or company that can issue or validate certificates. A trusted authority is sometimes called a "certificate authority."

Feedback Implosion

Feedback implosion (a multicast storm) is the overwhelming of a data network that occurs when multiple receivers provide feedback messages (such as acknowledgement messages) at approximately the same time. Feedback implosions may be avoided or reduced in severity through the use of feedback controls.

Reliable Multicast Transport Protocol (RMTP)

Reliable multicast transport protocol is a set of commands and processes that can coordinate the flow of data in a multicast tree. RMTP also has the capability to aggregate feedback acknowledgements at various distribution points within a multicast network to reduce the potential of feedback implosions.

Tree-Based Multicast Transport Protocol (TMTP)

Tree-based multicast transport protocol is a set of commands and processes that organize group members into domains where each domain contains a domain manager that is responsible for coordinating retransmissions that are used for error recovery.

Express Transport Protocol (XTP)

Express transport protocol is a set of commands and processes that are used to coordinate the transmission, reception and retransmission of packets in a data network to multiple recipients (multicast). XTP uses NACK messages to enable the reliable transfer of data in multicast distribution trees. XTP can utilize different types of network layer protocols (such as Internet Protocol).

Reliable Multicast Transport (RMT)

Reliable multicast transport is the processes that can be used to monitor and adjust to ensure that multicast data is delivered within defined performance requirements. Data reliability is the capability of a device or system to provide error detection, error recovery and/or loss protection.

Protocols and other device functions may be divided into building blocks that can be combined to enable key functions without all the implementation limitations of the protocols (such as limited scalability).

Reliability Mechanism

A reliability mechanism is the processes that are used to ensure the successful transmission and reception of data or media that is transferred through a network or device. Reliability is ensured through the use of redundancy and redundancy options in multicast systems include spatial redundancy and temporal redundancy.

Spatial redundancy is the process of providing related or repeated information within the transmission space. Forward error correction (FEC) is a form of spatial redundancy where some of the transmission resources are used to transfer redundant (error correction) information. Temporal redundancy is the repetition of information over a period of time. Repeated transmissions or retransmission requests and acknowledgements are forms of temporal redundancy.

Error Detection

Error detection is the process of detecting for bits that are received in error during data transmission. Error detection is made possible by sending additional bits that have a relationship to the original data that can be verified.

Error Recovery

Error recovery is the ability of a receiver or decoder to recover from catastrophic bit errors or lost bits within the bitstream. For example, a decoder may need to terminate decoding the previous bitstream segment and completely resynchronize with a new synchronization point with the bitstream. The speed of error recovery may depend on both bitstream characteristics and decoder implementation.

Error recovery processes in multicast systems can be centralized, distributed, or redirected to alternative packet repair nodes. Retransmission messages can be unicast or sent as multicast messages, which allows all multicast receivers to have access to the retransmitted information.

Error Protection

Error protection is the process of adding information to a data signal (typically by sending additional data bits) that permits a receiver of information to detect and/or correct errors that may have occurred during data transmission.

Error protection requires the use of additional transmission bandwidth for inclusion of error protection bits. Some transmitted bits are more important than others. Multicast systems can selectively add error protection to media streams using asynchronous layered coding (ALC).

Congestion Control

Congestion control is a process or set of processes that can be used to control the flow or processing of data in a system or network when the resources used in the system may exceed its capacity to deliver that service.

Congestion feedback is the providing of information that describes the over utilization or potential of over utilization of a network distribution system. Congestion feedback may be used to adjust transmitted rate control or to reconfigure receivers to reduce the congestion (over utilization) of the network

Feedback Control

Feedback controls are actions or processes taken by a device, system or network to regulate or adjust the timing and amount of feedback that devices or services create.

Real Time Feedback

Real time feedback is the ability of a sender to receive feedback information within a short time from when it was originated. Feedback timing in some systems (such as multicast systems) may be controlled or delayed to reduce the potential of feedback implosions.

Acknowledgement Implosion (Ack Implosion)

Acknowledgment implosion is the overwhelming of a data network that occurs when multiple receivers transmit acknowledgement messages at approximately the same time.

Crying Baby

A crying baby is a device or user that continuously requires attention or maintenance. If a crying baby is determined, it can be controlled or reconfigured to reduce the amount of messages or feedback data that it is sending.

Flow Control

Flow control is a combination of hardware or software mechanisms or protocols that can manage data transmissions when the receiving device cannot accept data at the same rate the sender is transmitting. Flow control is used when one of the devices communicating cannot receive the information at the same rate as it is being sent; this usually occurs when extensive processing is required by the receiver and the receive buffers are running low.

Ordering Guarantee

Ordering guarantee is the commitment or capability of a system to provide the delivery of packets in the same order that they were transferred into the network. Ordering guarantees may be assured by using higher level protocols that can sequentially identify and reorganize received packets into their original transmitted order.

Scalability

Scalability is the ability of a system to increase the number of users or amount of services it can provide without significant changes to the hardware or technology used. Multicast scalability is the ability to provide multicast distribution and control capabilities when the number of users and network elements is increased.

Late Join

Late join is the ability of a communication device to be added to a communication session after the communication session has been established. Late joining is also called grafting.

Multiple Passes

Multiple passes is the process of sending data multiple times to the same device. This allows the receiving device to capture and combine data from one pass with data from another pass to construct a complete or more complete packet or block of data.

Asynchronous Layered Coding (ALC)

Asynchronous layered coding is a forward error correction (FEC) transmission process that sends FEC information on multiple layers which allows receives to successfully recover information without the need for direct feedback.

Scalable Reliable Multicast (SRM)

Scalable reliable multicast is a media distribution system that provides group media distribution on a best effort basis. SRM uses multicast retransmission, local recovery and implosion avoidance techniques to ensure the multicast session can reliably operate even when there are many multicast group members.

Retransmissions are multicast which allows other recipients to receive corrective information without the need to send retransmission requests. Repair requests can be limited to local recovery groups to minimize network traffic. Other signaling control processes can be used to reduce the potential of the simultaneous sending of retransmission requests resulting in network implosion (such as the use of backoff timers).

Multicast Quality of Service (QoS)

Multicast quality of service (QoS) is one or more measurements of desired performance and priorities of a communications system. Multicast QoS may be managed or controlled through the use of bandwidth allocation, path precedence, resource reservation and class based controls.

Bandwidth Allocation

Bandwidth allocation is the frequency width of a radio channel in Hertz (high and low frequency limits) that can be modulated (changed) to transfer information (voice or data signals). The amount and type of information being sent determines the amount of bandwidth used and the method of modulation used to impose the information on the radio signal.

Retransmission

Retransmission is a method of network error control in which hosts receiving messages acknowledge the receipt of correct messages and either do not acknowledge or acknowledge in the negative the receipt of incorrect messages. The lack of acknowledgment, or receipt of negative acknowledgment, is an indication to the sending host that it should transmit the failed message again.

Path Precedence

Path precedence is categorization and prioritization of connection paths through a network or switching system.

IP Precedence utilizes the 3 precedence bits in the Type-of-Service (TOS) field in the IP header to specify class of service assignment for each IP packet. IP Precedence provides considerable flexibility for precedence assignment including customer assignment (e.g. by application) and network assignment based on IP or MAC address, physical port, or application. IP Precedence enables the network to act either in passive mode (accepting precedence assigned by the customer) or in active mode utilizing defined policies to either set or override the precedence assignment. IP Precedence can be mapped into adjacent technologies (e.g. Frame Relay or ATM) to deliver end-to-end QoS policies in a heterogeneous network environment. Thus, IP Precedence enables service classes to be established with no changes to existing applications and with no complicated network signaling requirements.

Resource Reservation

Resource reservation protocol is a set of commands and processes that can be used to control the quality of service (QoS) for packet flow for specific communication sessions in a packet data network. RSVP is primarily used in real-time communication sessions such as video or voice services over packet data networks.

Multicast resource reservation involves the requesting of resources, receiving acknowledgement that the reservation has been accepted, modified or rejected and releasing resources when users disconnect or the session has ended.

For multicast systems, resources are requested and assigned dynamically as trees are grown, leaves are added (new group recipients), and leaves are removed (users disconnect from the multicast session).

Service Classes

Service classes are sets of communication parameters that are used or assigned to provide transmission flows that can provide services that meet specific quality of service (QoS) requirements.

Class of service is the communication parameters that are assigned or associated with a particular application or communication session. The class of service usually requires a specific quality of service (QoS) level.

Class based queuing is a priority scheduling method that gives preferential treatment traffic communication sessions that have different priority levels for each class of communication session (such as real-time audio traffic and near-real time video). CBQ allocates variable bandwidth for each class of traffic so all applications continue to operate under heavy traffic conditions.

Guaranteed Service

Guaranteed service provides a specific bandwidth with a set maximum end-to-end transmission delay time. Guaranteed service may involve the allocation of dedicated resources (such as packet processing capacity in routers) that ensures that services are provided within agreed performance limits (such as maximum delay and error rates).

Controlled Service

A controlled service is the providing of information transfer, processes or authorizations that enable users, devices or systems to perform actions that they desire where the processes or operations are monitored and adjusted to meet desired criteria such as the maximum delay or the peak amount of data transmission.

Best Effort

Best effort is a level of service in a communications system that doesn't have a guaranteed level of quality of service (QoS).

Congestion Control

Congestion control is a process or set of processes that can be used to control the flow or processing of data in a system or network. Congestion control may be coordinated through the use of a packet scheduler. A packet scheduler is a device or function that can identify, prioritize, and coordinate a time that packets should be transmitted.

Admission Control

Admission control is the process of reviewing the service authorization level associated with users and determining how much network resources will be allocated if the network resources are available. Admission control is used to adjust, limit or assign the use of limited network resources to specific types or individual users. Access control may allow for the assignment of higher access level priority for specific types of users such as public safety users.

Gridcasting

Grid casting is the process of transmitting media channels to a number of users through the use of nodes that receive and redistribute (copying media channels) as they progress through a network.

Peercasting

Peercasting is the process of transmitting media channels to a number of users were media is received and redistributed by peers as it progresses through the network. Peercasting systems define how users can discover, connect and transmit from and to other peers within the network.

Figure 1.19 shows how a peercasting system can operate. This example shows that a peercasting system retransmits signals from one receiver to

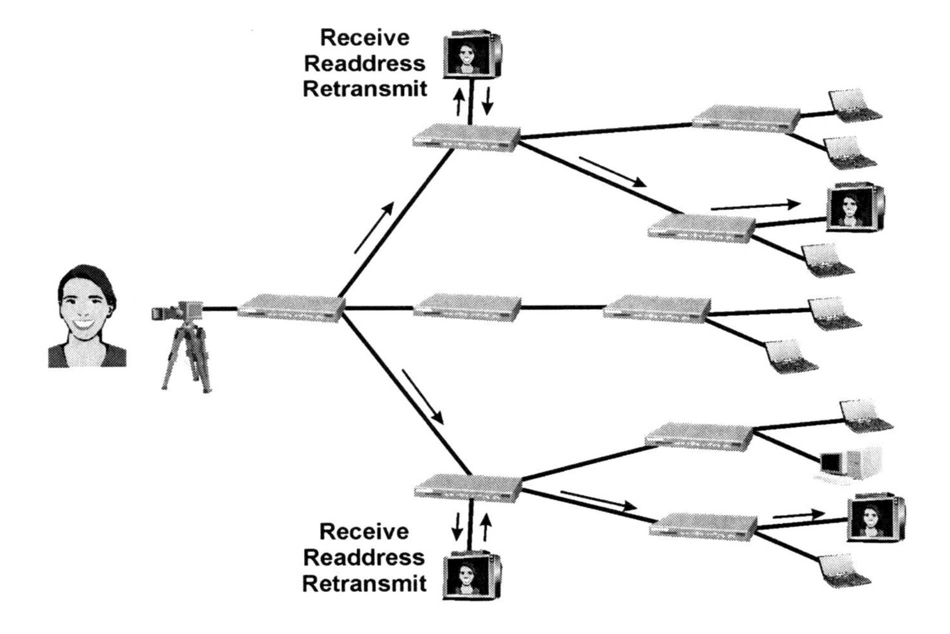

Figure 1.19, Peercasting Operation

other receivers. In this peercasting system, end user devices receive, readdress and retransmit received packets towards a new destination device.

Bit Torrent

A bit torrent is a rapid file transfer that occurs when multiple providers of information can combine their digital data (bits) transfer into a single stream (a torrent) of file information to the receiving computer.

Figure 1.20 shows how to transfer files using the torrent process. This example shows that 4 computers contain a large information file (such as a movie DVD). Each of the computers is connected to the Internet via high-speed connections that have high-speed download capability and medium-speed

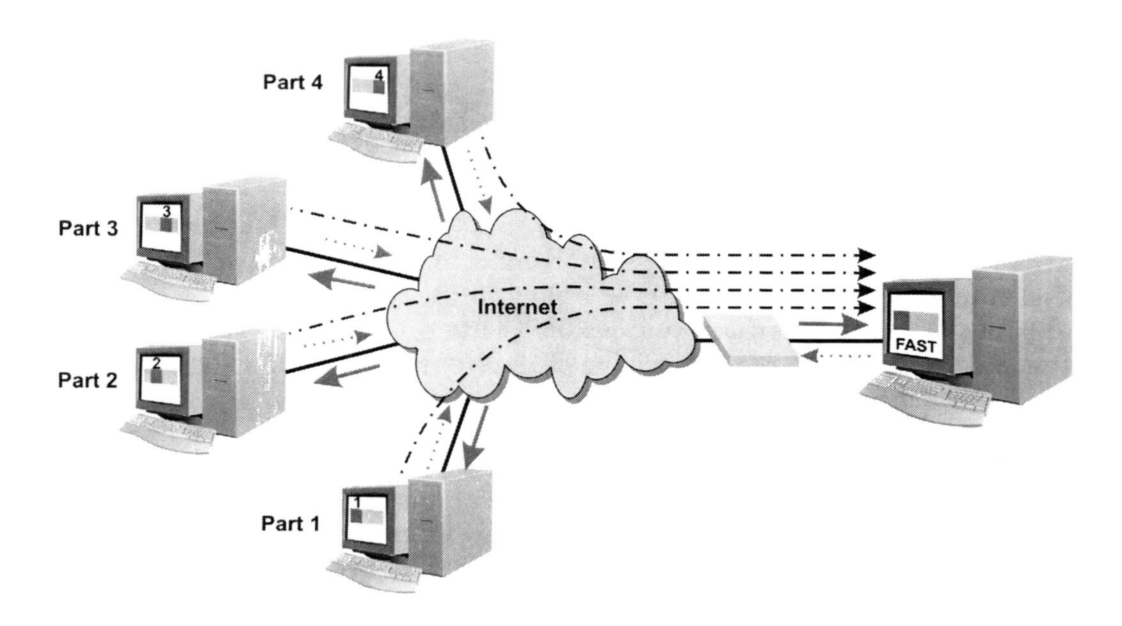

Figure 1.20, Bit Torrent

upload capability. To speed up the transfer speed for the file transfer, the receiver of information can request sections of the media file to be downloaded. Because the receiver of the information has a high-speed download connection, the limited uplink data rates of the section suppliers are combined. This allows the receiver of the information to transfer the entire file much faster.

Internet2 Network

Internet2 is a second generation of the Internet that uses a high-speed backbone communications network. The Internet system is a result of the next generation Internet (NGI) initiative that is sponsored by the United States government. Internet2 is seen as the way to deliver multimedia content (e.g. video on demand) through the Internet.

Multicast Backbone (MBone)

Multicast backbone (MBone) is a high-speed data communications system that interconnects with the Internet to provide multicast services.

Routers in the Internet are not setup to provide multicast services. The MBONE network is composed of interconnected networks that have multicast routing capability that can be linked to the Internet. Signals from one area of the network can be multicasted to other areas of the Internet. Once the multicast signals reach the edge of the Internet, local networks can distribute the multicast signals to their destination.

The MBone system may interconnect multicast routers through the use of MBone tunnels. MBone tunnels send MBone multicast packets through unicast routers that are not capable of or setup for processing multicast packets. The tunnel provides a point to point link between MBone multicast routers and the multicast packets are embedded in the unicast packet payload. When the unicast packet reaches its MBone multicast router destination, it is extracted from the payload and again processed as a multicast packet.

Figure 1.21 shows that the MBone is a multicast enabled network that can use multicasting to link multiple portions of the Internet with each other. The MBone system can be connected to routers in the Internet. The MBone contains multicast routers, which route the same signal to multiple portions of the Internet. The MBone is then connected back to the Internet where local routers can complete the multicast distribution.

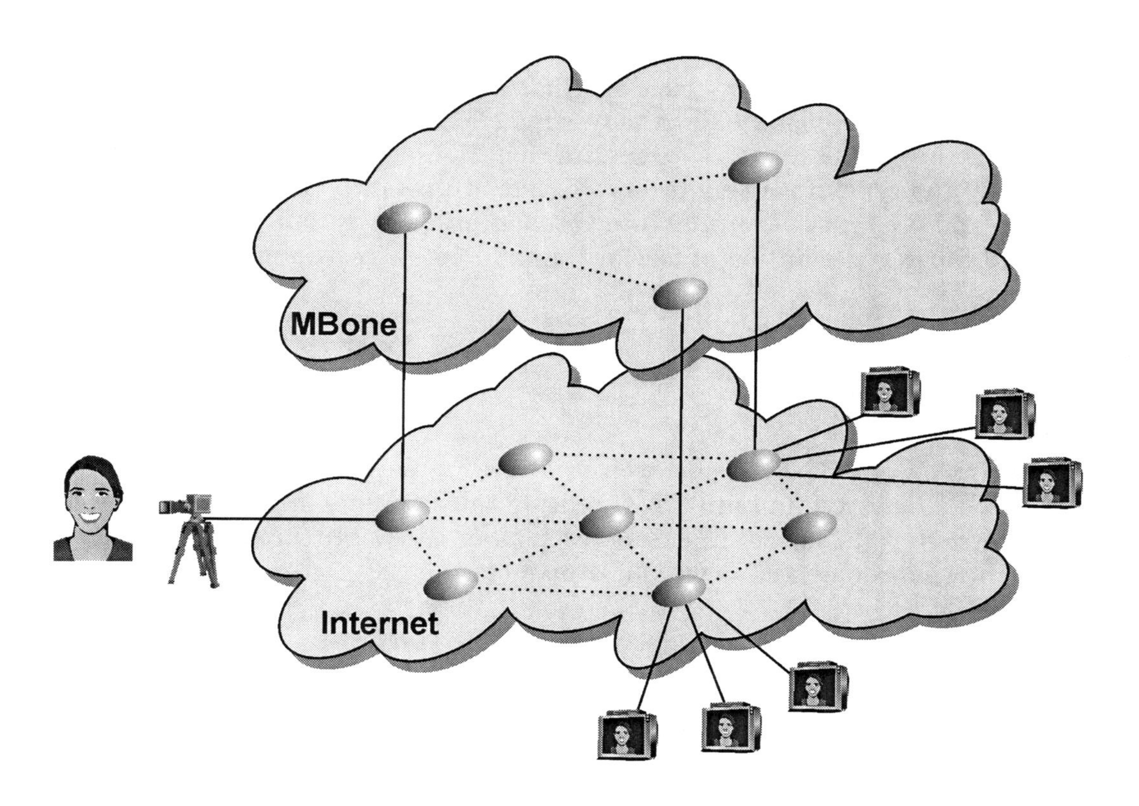

Figure 1.21, Multicast Backbone

Data Multicasting Constraints

Data multicasting constraints are limitations or tradeoffs that occur as multicast system options are selected. These limitations and options range from the capability of routers to perform multicast operations to the delays that occur when users change channels. Some of the key aspects to consider when setting up multicast systems include the maximum number of members, their geographic distribution, the type of application (real time or non-real time), the service data rates, how long the session will last and interactivity requirements.

Router Processing

Router processing is the instructions that a packet routing device performs to receive packets, lookup address routing choices in a routing table and to transmit the packet towards its destination. Router processing can be influenced by the type of routing protocols that are used (link state or distance vector) and the types of routing (unicast or multicast). Router processing can be used during calculation of forwarding addresses and updating of routing tables.

Router Memory

Router memory is the storage area in a routing device that holds information such as unicast and multicast routing tables. Some multicast protocols require routers to maintain extensive lists of tree structures and member state information, which uses router memory.

Multicast Setup Time

Multicast setup time is the duration between the initiation of a multicast service request (such as a join request) and when the service or action that is requested begins to operate. The setup time for a user to attach to a multicast stream is called the join latency. Join latency is the time duration between the start of a multicast join process (such as initiating a channel change) and when the group participant receives the media (when the new channel begins to display).

Protocol Overhead

Protocol overhead is the combined amount of data that messages consume compared to the user data or media that is transferred by a communication connection or service. Some multicast protocols use a higher percentage of bandwidth for signaling commands.

Transmission and Control Delays

Transmission and control delays are time intervals that occur between the sending of data or commands and when that data or responses to commands are received. Multicast system control delays such as the time when media flow control commands are initiated (such as mute or viewing pause) and when the operation is performed can reduce the user's quality of experience (QoE).

Reliability

Reliability is the ability of a network or equipment to perform within its normal operating parameters to provide a specific quality level of service. Reliability can be measured as a minimum performance rating over a specified interval of time. Multicast reliability can be provided by using a variety of forward error coding and retransmission options.

Media Heterogeneity

Media heterogeneity is the ability of a communication system to process or transmit media of different types such as digital video and digital audio. Distributing different types of media (audio and video) over the same multicast tree may result in high quality of one type of media (such as audio) while another type of media (such as video) may have poor quality levels.

Group Management

Group management is the process of defining groups of users or devices and adding and removing members (people and/or devices) to the groups. Group management may need to perform over several domain groups, control membership (access control), and provide tracking (usage) information about multicast sessions.

Quality of Service

Quality of service (QoS) is one or more measurement of desired performance and priorities of a communications system. QoS measures may include service availability, maximum bit error rate (BER), minimum committed bit rate (CBR) and other measurements that are used to ensure quality communications service.

Late Entry

Late entry is the ability of a communication device to be added to a communication session after the communication session has been established. Late entry is also known as grafting.

Scalability

Scalability is the ability of a system to increase the number of users or amount of services it can provide without significant changes to the hardware or technology used.

Security

Security is the ability of a person, system or service to maintain its desired well being or operation without damage, theft or compromise of its resources from unwanted people or events.

Figure 1.22 shows some of the capabilities and options that can be considered for multicast systems. The tradeoffs that may be considered when set-

Capabilities	Notes
Router Processing	The amount of instructions necessary for routers to determine if and how packets should be forwarded in the multicast tree.
Router Memory	Routing table size varies based on the selection of multicast protocols and the number of devices in the multicast tree.
Multicast Setup Time	How long it takes to setup multicast connections.
Protocol Overhead	Percentage of bandwidth needed for the signaling control commands.
Transmission and Control Delays	Time lag between changes (such as media channel changes).
Reliability	Ability of multicast sessions to continue when packets are corrupted or lost.
Media Heterogeneity	Ability to transmit media in different formats (e.g. audio and video).
Group Management	Controlling who can join multicast sessions and how they can join.
Quality of Service (QoS)	Ability to assign, manage, and provide different types of services.
Late Entry	Ability to dynamically add new group members after the multicast session has started.
Scalability	Ability to add or provide service to many group members. Potential for feedback implosion.
Security	Ability to restrict membership, assign and distribute keys to multiple group members, and encrypt (protect the streaming media).

Figure 1.22, Data Multicasting Constraints

ting up multicast systems and multicast protocol selection include router processing, router memory, channel setup time, command control times, multimedia (audio and video) transmission, media reliability, and other factors.

References and Resources

1. "IP Multicast", Shivkumar Kalyanaraman, Rensselaer Polytechnic Institute,
http://www.ecse.rpi.edu/Homepages/shivkuma/teaching/sp2000/i16_mult.ppt

2. "Internet Protocol (IP) Multicast", Cisco, http://www.cisco.com/warp/public/cc/pd/iosw/prodlit/ipimt_ov.pdf

3. "IP Multicasting, The Complete Guide to Interactive Corporate Networks," Dave Kosiur, ISBN: 0-471-24359-0, John Wiley, 1998.

4. "Host Extensions for IP Multicasting," Internet Engineering Task Force (IETF), http://www.ietf.org/rfc/rfc1112.txt.

5. "Protocol Independent Multicast-Sparse Mode (PIM-SM): Protocol Specification," Internet Engineering Task Force (IETF), http://www.ietf.org/rfc/rfc2362.txt.

6. "Reliable Multicast Transport Building Blocks for One-to-Many Bulk-Data Transfer," Internet Engineering Task Force (IETF), http://www.ietf.org/rfc/rfc3048.txt.

Appendix 1 - Acronyms

ABR-Area Border Router
Ack Implosion-Acknowledgement Implosion
AFDP-Adaptive File Distribution Protocol
ALF-Application Layer Framing
ARP-Address Resolution Protocol
AS-Autonomous System
ASBR-Autonomous System Boundary Router
ASM-Any Source Multicast
BCMCS-Broadcast and Multicast Services
BE-Best Effort Service
BGMP-Border Gateway Multicast Protocol
Bidir-PIM-Bidirectional Protocol Independent Multicasting
BMC-Broadcast/Multicast Control Protocol
BMDP-Broadcast Media Distribution Protocol
BMSC-Broadcast Multicast Service Center
BR-Border Router
CAM-Content Addressable Memory
CBQ-Class Based Queuing
CBT-Core Based Trees
CGMP-Cisco Group Management Protocol
CID-Conference Identifier
COS-Class Of Service
DCP-Dynamic Configuration Protocol
DVB-Data-Digital Video Broadcasting Data
DVMRP-Distance Vector Multicast Routing Protocol
EBCMCS-Enhanced Broadcast and Multicast Services
eBGP-External Border Gateway Protocol
EGP-Exterior Gateway Protocol
EIGRP-Enhanced Interior Gateway Routing Protocol
ERP-Exterior Routing Protocol
FC-Feedback Control
GM-Group Management
GMP-Group Membership Protocol
Gridcasting-Grid Casting
HDVMRP-Hierarchical Distance Vector Multicast Routing Protocol
HPIM-Hierarchical Protocol Independent Multicast
iBGP-Internal Border Gateway Protocol
ICMP-Internet Control Message Protocol
IGMP-Internet Group Management Protocol
IGMP Snooping-Internet Group Management Protocol Snooping
IGMPv1-Internet Group Management Protocol Version 1
IGMPv2-Internet Group Management Protocol Version 2
IGMPv3-Internet Group Management Protocol Version 3

IGP-Interior Gateway Protocol
IGRP-Interior Gateway Routing Protocol
InPort-Input Port
Intserv-Integrated Services
Intserv)1-Integrated Services
IP Address-Internet Protocol Address
IP Subnet-Internet Protocol Subnetwork
IPv6-Internet Protocol Version 6
IPVBI-IP Multicast over VBI
IPX-Internetwork Packet Exchange
IR Routing-Infrared Routing
Keepalive-Keep Alive Message
LBRM-Log Based Receiver Reliable Multicast
LDP-Label Distribution Protocol
LGM-Leave Group Message
LGMP-Local Group Multicast Protocol
LSA-Link State Advertisement
LSP-Label Switched Path
LSR-Label Switched Router
M-Merging Point
M2M-Many to Many
MAAP-Multicast Address Allocation Protocol
MARS-Multicast Address Resolution Server
MBGP-Multicast Border Gateway Protocol
M-BGP-Multicast Border Gateway Protocol
MBMS-Multimedia Broadcast Multicast Services
MBONE-Multicast Backbone
MBR-Multicast Border Router
MCP-Multicast Control Protocol
MCU-Multipoint Control Unit
MFC-Multicast Forwarding Cache
MFTP-Multicast File Transfer Protocol
MGM-Multicast Group Manager
M-IGP-Multicast Interior Gateway Protocol
MLD-Multicast Listener Discovery
MOSPF-Multicast Extensions to Open Shortest Path First
MPE-Multiprotocol Encapsulation
MPLS-Multi-Protocol Label Switching
M-RIB-Multicast Routing Information Base
MRouter-Multicast Router
MSDP-Multicast Source Discovery Protocol
MTFTP-Multicast Trivial File Transfer Protocol
MTP-Multicast Transport Protocol
MTP-2-Multicast Transport Protocol Version 2
Multicast CIDs-Multicast Polling Connection Identifiers
NACK-Negative-Acknowledgment
NACK Implosion-Negative Acknowledgement Implosion
NTP-Network Time Protocol
Outport-Output Port
PGM-Pragmatic General Multicast
Piggyback-Piggy-Back
PIM-Protocol Independent Multicast
PIM-DM-Protocol Independent Mulitcast-Dense Mode
PIM-SM-Protocol Independent-Sparse Mode
PKE-Public Key Encryption
PNAP-Private Network Access Point
PR-Bit-Poison Reverse Bit
QoS-Quality Of Service
QoS Awareness-Quality of Service Awareness
RAMP-Reliable Adaptive Multicast Protocol
RARP-Reverse Address Resolution Protocol
RBP-Reliable Broadcast Protocol

RGMP-Router-Port Group Management Protocol
RIP-Routing Information Protocol
RMF-Reliable Multicast Framework
RMFP-Reliable Multicast Framing Protocol
RMP-Reliable Multicast Protocol
RMT-Reliable Multicast Transport
RMTP-Reliable Multicast Transport Protocol
RP-Rendezvous Point
RPB-Reverse Path Broadcasting
RPF-Reverse Path Forwarding
RPM-Reverse Path Multicasting
RSVP-Resource Reservation Protocol
RTCP-Real-Time Transport Control Protocol
RTP-Reliable Transport Protocol
RTSP-Real Time Streaming Protocol
S,G-Source and Group Pair
SA-Security Association
SA-Source Active
SAP-Session Announcement Protocol
SCCP-Session Conference Control Protocol
SCE-Single Connector Emulation
SCP-Session Control Protocol
SDES-Source Description Packet
SDP-Session Description Protocol
SGMP-Source Filtering Group Membership Protocol
SIP-Session Initiation Protocol
SLP-Service Location Protocol
SNA-System Network Architecture
SPT-Shortest Path Tree
SPT Bit-Shortest Path Tree Bit
SRM-Scalable Reliable Multicast
SSM-Single Source Multicast
SSM-Single Source Multicasting
SSM-Source Specific Multicast
ST-II-Stream Protocol Version II
Subnet-Sub Network
TMTP-Tree-Based Multicast Transport Protocol
TP0-Transport Protocol Class 0
TP1-Transport Protocol Class 1
TP2-Transport Protocol Class 2
TP3-Transport Protocol Class 3
TP4-Transport Protocol Class 4
TRBP-Truncated Reverse Path Broadcasting
TRPB-Truncated Reverse Path Broadcasting
TTL-Time To Live
TTL Scoping-Time to Live Scoping
U-RIB-Unicast Routing Information Base
VC-Virtual Circuit
XTP-Xpress Transport Protocol

Index

Althos Publishing Book List

Product ID	Title	# Pages	ISBN	Price	Copyright
	Billing				
BK7781338	Billing Dictionary	644	1932813381	$39.99	2006
BK7781339	Creating RFPs for Billing Systems	94	193281339X	$19.99	2007
BK7781373	Introduction to IPTV Billing	60	193281373X	$14.99	2006
BK7781384	Introduction to Telecom Billing, 2nd Edition	68	1932813845	$19.99	2007
BK7781343	Introduction to Utility Billing	92	1932813438	$19.99	2007
BK7769438	Introduction to Wireless Billing	44	097469438X	$14.99	2004
	IP Telephony				
BK7781361	Tehrani's IP Telephony Dictionary, 2nd Edition	628	1932813616	$39.99	2005
BK7781311	Creating RFPs for IP Telephony Communication Systems	86	193281311X	$19.99	2004
BK7780530	Internet Telephone Basics	224	0972805303	$29.99	2003
BK7727877	Introduction to IP Telephony, 2nd Edition	112	0974278777	$19.99	2006
BK7780538	Introduction to SIP IP Telephony Systems	144	0972805389	$14.99	2003
BK7769430	Introduction to SS7 and IP	56	0974694304	$12.99	2004
BK7781309	IP Telephony Basics	324	1932813098	$34.99	2004
BK7780532	Voice over Data Networks for Managers	348	097280532X	$49.99	2003
	IP Television				
BK7781334	IPTV Dictionary	652	1932813349	$39.99	2006
BK7781362	Creating RFPs for IP Television Systems	86	1932813624	$19.99	2007
BK7781355	Introduction to Data Multicasting	68	1932813551	$19.99	2006
BK7781340	Introduction to Digital Rights Management (DRM)	84	1932813403	$19.99	2006
BK7781351	Introduction to IP Audio	64	1932813519	$19.99	2006
BK7781335	Introduction to IP Television	104	1932813357	$19.99	2006
BK7781341	Introduction to IP Video	88	1932813411	$19.99	2006
BK7781352	Introduction to Mobile Video	68	1932813527	$19.99	2006
BK7781353	Introduction to MPEG	72	1932813535	$19.99	2006
BK7781342	Introduction to Premises Distribution Networks (PDN)	68	193281342X	$19.99	2006
BK7781357	IP Television Directory	154	1932813578	$89.99	2007
BK7781356	IPTV Basics	308	193281356X	$39.99	2007
BK7781389	IPTV Business Opportunities	232	1932813896	$24.99	2007
	Legal and Regulatory				
BK7781378	Not so Patently Obvious	224	1932813780	$39.99	2006
BK7780533	Patent or Perish	220	0972805338	$39.95	2003
BK7769433	Practical Patent Strategies Used by Successful Companies	48	0974694339	$14.99	2003
BK7781332	Strategic Patent Planning for Software Companies	58	1932813322	$14.99	2004
	Telecom				
BK7781316	Telecom Dictionary	744	1932813160	$39.99	2006
BK7781313	ATM Basics	156	1932813136	$29.99	2004
BK7781345	Introduction to Digital Subscriber Line (DSL)	72	1932813454	$14.99	2005
BK7727872	Introduction to Private Telephone Systems 2nd Edition	86	0974278726	$14.99	2005
BK7727876	Introduction to Public Switched Telephone 2nd Edition	54	0974278769	$14.99	2005
BK7781302	Introduction to SS7	138	1932813020	$19.99	2004
BK7781315	Introduction to Switching Systems	92	1932813152	$19.99	2007
BK7781314	Introduction to Telecom Signaling	88	1932813144	$19.99	2007
BK7727870	Introduction to Transmission Systems	52	097427870X	$14.99	2004
BK7780537	SS7 Basics, 3rd Edition	276	0972805370	$34.99	2003
BK7780535	Telecom Basics, 3rd Edition	354	0972805354	$29.99	2003
BK7780539	Telecom Systems	384	0972805397	$39.99	2006

For a complete list please visit
www.AlthosBooks.com

ALTHOS

Althos Publishing Book List

Product ID	Title	# Pages	ISBN	Price	Copyright
Wireless					
BK7769431	Wireless Dictionary	670	0974694312	$39.99	2005
BK7769434	Introduction to 802.11 Wireless LAN (WLAN)	62	0974694347	$14.99	2004
BK7781374	Introduction to 802.16 WiMax	116	1932813748	$19.99	2006
BK7781307	Introduction to Analog Cellular	84	1932813071	$19.99	2006
BK7769435	Introduction to Bluetooth	60	0974694355	$14.99	2004
BK7781305	Introduction to Code Division Multiple Access (CDMA)	100	1932813055	$14.99	2004
BK7781308	Introduction to EVDO	84	193281308X	$14.99	2004
BK7781306	Introduction to GPRS and EDGE	98	1932813063	$14.99	2004
BK7781370	Introduction to Global Positioning System (GPS)	92	1932813705	$19.99	2007
BK7781304	Introduction to GSM	110	1932813047	$14.99	2004
BK7781391	Introduction to HSPDA	88	1932813918	$19.99	2007
BK7781390	Introduction to IP Multimedia Subsystem (IMS)	116	193281390X	$19.99	2006
BK7769439	Introduction to Mobile Data	62	0974694398	$14.99	2005
BK7769432	Introduction to Mobile Telephone Systems	48	0974694320	$10.99	2003
BK7769437	Introduction to Paging Systems	42	0974694371	$14.99	2004
BK7769436	Introduction to Private Land Mobile Radio	52	0974694363	$14.99	2004
BK7727878	Introduction to Satellite Systems	72	0974278785	$14.99	2005
BK7781312	Introduction to WCDMA	112	1932813128	$14.99	2004
BK7727879	Introduction to Wireless Systems, 2nd Edition	76	0974278793	$19.99	2006
BK7781337	Mobile Systems	468	1932813373	$39.99	2007
BK7780534	Wireless Systems	536	0972805346	$34.99	2004
BK7781303	Wireless Technology Basics	50	1932813039	$12.99	2004
Optical					
BK7781365	Optical Dictionary	712	1932813659	$39.99	2007
BK7781386	Fiber Optic Basics	316	1932813861	$34.99	2006
BK7781329	Introduction to Optical Communication	132	1932813292	$14.99	2006
Marketing					
BK7781323	Web Marketing Dictionary	688	1932813233	$39.99	2007
BK7781318	Introduction to eMail Marketing	88	1932813187	$19.99	2007
BK7781322	Introduction to Internet AdWord Marketing	92	1932813225	$19.99	2007
BK7781320	Introduction to Internet Affiliate Marketing	88	1932813209	$19.99	2007
BK7781317	Introduction to Internet Marketing	104	1932813292	$19.99	2006
BK7781317	Introduction to Search Engine Optimization (SEO)	84	1932813179	$19.99	2007
Programming					
BK7781300	Introduction to xHTML:	58	1932813004	$14.99	2004
BK7727875	Wireless Markup Language (WML)	287	0974278750	$34.99	2003
Datacom					
BK7781331	Datacom Basics	324	1932813314	$39.99	2007
BK7781355	Introduction to Data Multicasting	104	1932813551	$19.99	
BK7727873	Introduction to Data Networks, 2nd Edition	64	0974278734	$19.99	2006
Cable Television					
BK7781371	Cable Television Dictionary	628	1932813713	$39.99	2007
BK7780536	Introduction to Cable Television, 2nd Edition	96	0972805362	$19.99	2006
BK7781380	Introduction to DOCSIS	104	1932813802	$19.99	2007
Business					
BK7781368	Career Coach	92	1932813683	$14.99	2006
BK7781359	How to Get Private Business Loans	56	1932813594	$14.99	2005
BK7781369	Sales Representative Agreements	96	1932813691	$19.99	2007
BK7781364	Efficient Selling	156	1932813640	$24.99	2007

CPSIA information can be obtained at www.ICGtesting.com
231602LV00003B/35/P

9 781932 813555